PHOTOSYNTHETIC LIFE

OXFORD BIOLOGY PRIMERS

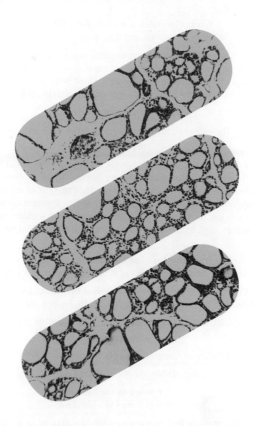

PHOTOSYNTHETIC LIFE
ORIGIN, EVOLUTION, AND FUTURE

Denis Murphy and Tanai Cardona

OXFORD
UNIVERSITY PRESS

OXFORD
UNIVERSITY PRESS

Great Clarendon Street, Oxford, OX2 6DP,
United Kingdom

Oxford University Press is a department of the University of Oxford.
It furthers the University's objective of excellence in research, scholarship,
and education by publishing worldwide. Oxford is a registered trade mark of
Oxford University Press in the UK and in certain other countries

© Oxford University Press 2022

The moral rights of the authors have been asserted

Impression: 1

Published in the United States of America by Oxford University Press
198 Madison Avenue, New York, NY 10016, United States of America

British Library Cataloguing in Publication Data

Data available

Library of Congress Control Number: 2022938278

ISBN 978-0-19-881572-3

Printed in the UK by
Bell & Bain Ltd., Glasgow

Links to third party websites are provided by Oxford in good faith and
for information only. Oxford disclaims any responsibility for the materials
contained in any third party website referenced in this work.

TABLE OF CONTENTS

ABBREVIATIONS (SEE ALSO GLOSSARY)

ATP	adenosine triphosphate
CAM	crassulacean acid metabolism
CASH lineage	cryptophyte, alveolate, stramenopile, and haptophyte (CASH) lineage
CEF	cyclic electron flow
CETCH	crotonyl-CoA/ethylmalonyl-CoA/hydroxybutyryl-CoA
ER	endoplasmic reticulum
Ga	giga annum (billion years)
Gb	giga-base pairs of DNA
GOE1	1st great oxygenation event
GOE2	2nd great oxygenation event
Ha	hectare
HGT	horizontal gene transfer
Hlip	high-light-inducible, single transmembrane helix protein
Kb	kilo-base pairs of DNA
KT	cretaceous –tertiary
LBCA	last bacterial common ancestor
LCHII	light harvesting complex II
LCHII	light harvesting complex I
LECA	last eukaryotic common ancestor
LEF	linear electron flow
LHC	light harvesting complex
LUCA	last universal common ancestor
Ma	mega annum (million years)
Mb	mega-base pairs of DNA
MEP	methylerythritol phosphate
Mha	million hectares
MOG	malonyl-CoA-oxaloacetate-glyoxylate
MRCA	most recent common ancestor
NADPH	reduced nicotinamide adenine dinucleotide phosphate
nm	nanometres
NOE	neoproterozoic oxygenation event
OCP	orange carotenoid protein
PSI	photosystem I
PSII	photosystem II
RC	reaction centre
ROS	reactive oxygen species
RPP	reductive pentose phosphate
Rubisco	ribulose-1,5-bisphosphate carboxylase/oxygenase
SOD	superoxide dismutase
SYM	symbiosis
t	tonne

FOREWORD

This book is about the origins and development of photosynthetic life on Earth. We start our story over four billion years ago when living organisms fist evolved and finish by examining its future prospects in a rapidly changing world. We follow the momentous consequences of **photosynthesis** on both biological and non-biological evolution on a global scale. For example, in terms of biology, photosynthesis led to the evolution of eukaryotes and hence all complex life forms, including **plants** and animals. It has also drastically altered the wider environment of our planet, including its climate, geochemistry, and the composition of its atmosphere. We have been careful to cover the entirety of the story of photosynthetic life, rather than just focusing on the higher plants. Indeed, the land plants have only been in existence for less than 15% of the time that bacteria and algae were already photosynthesizing and evolving on Earth.

We are living at an exciting time of great advances in our understanding of photosynthesis and evolution. Over recent years, new research tools have been applied from disciplines such as biology, chemistry, physics, geology, computation, and mathematics. This has contributed to a much more detailed understanding of the processes leading to cellular life, which started shortly after the formation of the Earth about 4.3 billion years ago. The history of research into photosynthesis is well documented elsewhere. Instead, our focus here is very much on our contemporary understanding of the evolution of photosynthetic life. We also look to the future and how we might be able to engineer more efficient versions of photosynthesis to address major global challenges, such as food security and climate change.

The rapid pace of recent research in this field is demonstrated in the references and suggestions for further reading, most of which were published during the decade up to the time of writing of the book in 2021. In this dynamic field of science, some of the more recent findings have yet to be fully confirmed but, if correct, they could fundamentally alter our view of photosynthesis as well as giving new perspectives on plant evolution. For this reason, we have tried to present a 'snapshot' of our current understanding that includes discussion of some areas that are still controversial.

Some of the principal aspects of photosynthesis that will be examined here include:

- the mechanisms of the various forms of photosynthesis
- where and how photosynthesis might have evolved
- how it was spread to so many groups of organisms
- how it changed the geochemistry of the Earth
- how it changed the trajectory of biological evolution
- how it enabled life to move from the seas to the land
- how it has been impacted by recent human activities
- how it might be manipulated to address future challenges

The book is aimed principally at advanced undergraduate and postgraduate students and others with interests in biology and evolution. It would also be useful for researchers in related disciplines, such as environmental sciences, who wish to update their knowledge of this important process. Each chapter includes text boxes that examine selected key topics, plus a list of further reading materials, most of which are available online. At the end of the book, there is a detailed glossary of technical terms used in the various chapters.

1 PHOTOSYNTHESIS, OXYGEN, AND THE EVOLUTION OF LIFE

Learning objectives

- Understanding the basic processes of photosynthesis
- How photosynthesis relates to the origin of life on Earth
- Tracing the origins of the first **prokaryotes**
- Tracing the origins of the first **eukaryotes**
- Following the global impacts of photosynthesis on life, the atmosphere, and geology of Earth following the establishment of an oxygenated planet

1.1 Introduction

This book is about how photosynthesis evolved on our planet. Photosynthesis is arguably the most important biological process on Earth and the question of what constitutes photosynthetic life is discussed in the Bigger picture 1.1. Either directly, as in plants and algae, or indirectly, as in animals and other heterotrophs, photosynthesis sustains almost all living organisms. It does this by using energy derived from sunlight to build up complex organic molecules, such as carbohydrates, proteins, and nucleic acids. Photosynthesis is also directly responsible for creating and maintaining the oxygen-rich atmosphere that is essential for all aerobic life. And finally, photosynthesis has played a key role in a multitude of abiotic (non-biological) geological processes that have shaped the composition of our planet as we know it today.

Photosynthesis first arose in simple unicellular bacteria early in the evolution of life, and the photosynthetic cyanobacteria are still important components of many global ecosystems today. About two billion years ago, or 2.0 Ga (giga annum or Ga), a cyanobacterium was taken up by a eukaryotic cell to form a unique symbiotic relationship that created the algae, or phytoplankton, that are the basis of marine food webs. The descendants of these symbiotic cyanobacteria are still present in all photosynthetic eukaryotic cells in the form of chloroplast organelles. Eventually, one group

Bigger picture 1.1
Defining photosynthetic organisms

Photosynthetic organisms include bacteria, algae, and land plants. The term plant has been used historically to mean the mainly terrestrial, multicellular, autotrophs that visually dominate most land-based ecosystems. In contrast, algae are simpler, often unicellular, photosynthetic organisms found in aquatic ecosystems. Both plants and algae are eukaryotes. The third major group of photosynthetic organisms is made up from several groups of bacteria. These are unicellular or colonial prokaryotes that inhabit almost all ecosystems on and under the Earth.

Here, we define a plant as a member of the Embryophytes, a group that is also known as the land plants—although some species have secondarily readopted aquatic lifestyles. The Embryophytes and their algal relatives, the Charophytes, make up the Streptophytes which, together with the Chlorophytes, in turn make up the Viridiplantae. The term, Viridiplantae, means 'green plants' and encompasses all of the green algae and land plant species.

The Viridiplantae, plus the red algae and glaucophytes, make up the Archaeplastida, which includes all the descendants of the primary endosymbiotic event whereby a cyanobacterium was captured by a heterotrophic eukaryotic cell and transformed into a plastid organelle.

Many other groups of so-called secondary algae, including diatoms and dinoflagellates, are the result of secondary endosymbioses between a wide range of eukaryotic hosts and red and/or green algae that were ingested and converted into so-called secondary plastids.

Finally, there are numerous groups of photosynthetic bacteria. Only one of these, the cyanobacteria, carries out a similar form of oxygenic photosynthesis to that found in plants and algae. The other photosynthetic bacteria perform various types of anoxygenic photosynthesis, which do not involve either water-splitting or oxygen evolution.

Although all the above groups are mostly made up of photosynthetic species, there are numerous instances where some species have evolved non-photosynthetic lifestyles. For example, several groups of land plants have become parasitic and have either partially or totally lost their ability to photosynthesize. The same is true for many algal species that have evolved into a wide variety of either parasitic or free-living heterotrophs.

This means that we cannot automatically equate all plants or algae with photosynthesis, and neither can we consider all the algae as a single group Instead, we have taken an evolutionary approach here. This helps us to understand both the links and the differences between the many diverse groups of organisms that are, or once were, capable of photosynthesis.

of green algae evolved into more sophisticated multicellular forms that became the first true land plants. Together with animals and fungi, land plants are the most common and familiar macroscopic organisms in contemporary terrestrial ecosystems.

During the past decade there have been several important advances in our knowledge of how and when life first evolved on Earth. The first living organisms were simple cells that possibly arose as early as 4 Ga. This happened soon after the planet had cooled sufficiently to enable oceans of liquid water to form. Recent genomic and geological studies suggest that photosynthesis is an ancient process that might have evolved shortly after the origin of the earliest cellular life. Photosynthesis then played a key role in the evolution of all life on Earth up to the present time. Photosynthesis has also changed the composition of many rock formations. One example is its role in the conversion of the widespread deposits of free metallic iron around the Earth into oxidized forms with very different geochemical properties. These iron ore deposits are now the source of metals for the iron and steel industries. Photosynthesis has also transformed the atmosphere from an anaerobic (oxygen-free) to an aerobic (oxygen-containing) mixture of gases able to sustain larger and more complex oxygen-requiring life forms, including humans.

As shown in Fig 1.1, in terms of total biomass, today's biosphere is dominated by just a few groups of photosynthetic organisms. The three most important are plants, algae, and cyanobacteria. It is estimated that these three groups of photosynthetic organisms collectively make up an impressive 85% of the entire terrestrial and aquatic biomass. In contrast, all of the animals (terrestrial and aquatic) collectively make up a mere 0.37% of total global biomass.

Fig. 1.1 Photosynthetic organisms dominate global biomass.

Estimated relative biomasses of the most abundant groups of contemporary organisms. Photosynthetic organisms are dominated by plants and algae but also include some aquatic protists and bacteria. Gt C, gigatonnes of carbon. In terms of cultivated/managed organisms our crop plants have 10-fold more biomass than our livestock and 17-fold more than humans.

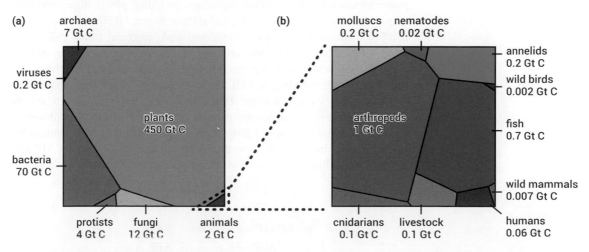

To put it another way, in terms of their total biomass, the three major groups of photosynthetic organisms are over 200 times more abundant than all marine and terrestrial animals put together.

In this chapter, we will briefly survey the basic processes involved in the various types of photosynthesis found in bacteria and eukaryotes and the pivotal role played by photosynthesis in the evolution of life on Earth. We will also consider the broader significance of photosynthesis in terms of the many important roles that it plays in both the biosphere and the geosphere.

1.2   What is photosynthesis?

The word 'photosynthesis' is derived from two Greek words meaning 'light' (*photos*) and 'putting together' (*synthesis*). In its simplest sense, photosynthesis can be defined as the use of light energy to enable the formation of complex organic molecules from relatively simple inorganic precursors. For example, most land plants use energy from visible sunlight (wavelengths from about 400 to 700 nm) to oxidize water molecules (H_2O) into protons (H^+) and electrons (e^-), with the release of oxygen gas (O_2) as shown in equation [1]. This particular form of photosynthesis is called *oxygenic*, or oxygen-producing, and is found in all plants and algae as well as in the cyanobacteria. Oxygenic photosynthesis is by far the most common and important process for the biological production of fixed organic carbon on Earth.

In oxygenic photosynthesis, light energy is first harvested by large pigment-protein complexes that are typically attached to thylakoid membranes located inside the chloroplasts of photosynthetic tissues, such as leaves. The energy is then channelled to other pigment-protein complexes called photochemical reaction centres (RCs) or photosystems. Here, chlorophyll molecules transform the energy of light into chemical energy, and electron flow, that is then used to power the oxidation of water. The protons and electrons from the splitting of water are used in subsequent reactions to power metabolism. All oxygenic organisms, including plants, algae, and cyanobacteria, rely on two types of RC, called Photosystem I (PSI) and Photosystem II (PSII) distinguished by their functional and structural traits. PSII is the water-splitting (water-oxidizing) protein complex of oxygenic photosynthesis.

As shown below in equations [1] and [2], protons and electrons released by water splitting in PSII are harnessed by other thylakoid membrane proteins via an electron transport chain that generates relatively stable high-energy compounds, such as ATP and NADPH. The processes involved in a) harvesting of light energy, b) its use to split water molecules in PSII, and c) the use of the resulting chemical energy to generate ATP and NADPH, are collectively known as the light reactions of photosynthesis.

$$H_2O \quad \xrightarrow{\text{sunlight}} \quad 2H^+ + e^- + O_2$$
$$NADP + H^+ + e^- \quad \text{>>>>>>} \quad NADPH$$
$$ADP + Pi \quad \text{>>>>>>} \quad ATP$$

Equation [1]. Highly simplified outline of the light reactions of oxygenic photosynthesis. In line 1, sunlight is harvested and used to split water with release of protons and electrons. In line 2, the released protons and electrons combine with NADP to form NADPH, a powerful reducing agent able to convert to sugars. In line 3, some of the energy is also used to produce ATP, which is the most common 'energy currency' used in cells.

The ATP and NADPH created by the light reactions provide the chemical energy and reducing ability that enables plants to convert atmospheric CO_2 plus water into simple sugars ($[CH_2O]_n$), such as glucose and sucrose. These simple sugars can be further elaborated to generate more complex compounds, such as the carbohydrates, amino acids, lipids, and nucleic acids that are the basis of cellular metabolism in all organisms.

$$
\begin{aligned}
& \qquad\qquad \text{ATP \& NADPH} \\
& CO_2 + H_2O \quad >>>>>>> \quad \text{Simple}[CH_2O]_n \\
& \text{Simple}[CH_2O]_n \quad >>>>>>> \quad \text{Complex}[CH_2O]_n + \text{proteins, DNA etc}
\end{aligned}
$$

Equation [2]. Highly simplified outline of photosynthetic carbon assimilation. In line 1, ATP and NADPH provide energy and reducing power to convert CO_2 to simple sugars such as glucose and sucrose. In line 2, ATP and other intermediates enable the simple sugars to be converted into more complex carbohydrates such as cellulose and starch as well as other essential metabolic end products such as proteins, lipids and DNA.

The overall process of oxygenic photosynthesis in terms of the light-driven reduction of CO_2 to carbohydrates via water splitting is summarized in equation [3] as follows:

$$
\begin{aligned}
& \qquad\qquad\quad \textit{light} \\
& 2H_2O + CO_2 \quad >>>>>> \quad [CH_2O]_n + H_2O + O_2 \\
& \qquad\quad 4e^-
\end{aligned}
$$

Equation [3]. Basic summary of oxygenic photosynthesis. Water is split by light energy leading to CO_2 reduction to carbohydrates and O_2 evolution.

The second form of photosynthesis is known as anoxygenic photosynthesis and is exclusively found within several groups of bacteria. As its name implies, anoxygenic photosynthesis occurs without the release of oxygen. Oxygenic photosynthesis is by far the most prevalent and productive photosynthetic process, accounting for the production of more than 99% of all organic matter in the planet. However, anoxygenic photosynthesis is also widespread in most illuminated environments, both marine and terrestrial, but only if they are anaerobic. While it is of local ecological relevance for its impact on the cycling of carbon, nitrogen, sulphur, and other nutrients, anoxygenic photosynthesis does not play a significant part in global biogeochemical cycles. Instead, these are dominated by oxygenic photosynthesis which is responsible for an estimated 3300-fold greater amount of carbon fixation (3,000 teramoles/year) compared to anoxygenic photosynthesis (2.7 teramoles/year).

During anoxygenic photosynthesis, water is replaced as the electron donor by substances such as hydrogen (H_2), hydrogen sulphide (H_2S), or iron (Fe^{2+}) as the electron source. Anoxygenic photosynthetic bacteria specialize in the use of far-red and near-infrared light using wavelengths between

Scientific approach 1.1
Structural studies of photochemical reaction centres

During recent years, the structures of all types of reaction centres (RCs) from both bacteria and photosynthetic eukaryotes have been characterized at an unprecedented level of atomic detail. The first solved structure of a RC, published in 1985, was also the first structure solved of any membrane protein. This was the structure of an anoxygenic type II RC of the proteobacterium *Blastochloris* (*Rhodopseudomonas*) *viridis* and this achievement won three researchers the 1988 Nobel Prize in chemistry. It was not until the early 2000s that the crystal structures of PSI and PSII from a thermophilic cyanobacterium were published following the efforts of several international teams.

In 2011, an impressive 1.9 Å resolution crystal structure of PSII from the cyanobacterium *Thermosynechococcus vulcanus* was published by a group of Japanese scientists. This showed for the first time the position of all atoms in the unique water-splitting complex that is made of four manganese ions bound by five oxygen-bridging atoms and one calcium in what looks like a distorted 'chair' configuration (see Chapters 2 and 3 for further details). At that time, that was the highest resolution structure achieved for any membrane protein complex.

Since 2011 there have been further technical advances in the resolution potential of cryo-electron microscopy and crystallography equipment. This has allowed researchers access to the structures of the major photosystem super-complexes from a wide variety of algae and plants. These are some of the largest molecular machines known in biology, with each complex harbouring hundreds of protein subunits and thousands of pigment molecules.

Of particular importance is the development of a sophisticated technique known as serial femtosecond crystallography using X-ray free electron lasers. This highly advanced method produces movies at atomic detail of enzymes as they progress through their catalytic cycles under native conditions. It has permitted several international teams of scientists to generate a complete visualization in PSII of every atom in the water-splitting complex as it progresses through the oxygen evolution cycle. Less than ten years ago such a feat would have been considered an impossible dream worthy of science fiction.

Looking further to the future, artificial intelligence, or AI, will play an increasingly prominent role in structural studies of proteins. In 2021, an AI company, called DeepMind, announced the use of their AlphaFold algorithm to solve the structure of over 365,000 proteins including many photosynthetic RC components. This new structural database will be of great benefit to researchers in the coming years.

800 to nearly 1000 nm to drive their photochemical RCs. To harvest the lower energy light radiation, anoxygenic photosynthetic bacteria use bacteriochlorophylls, which are related to chlorophylls, and are able to absorb longer wavelengths of light beyond the visible range. Their reliance on less efficient light harvesting and photochemical processes means that anoxygenic bacteria grow more slowly and, being anaerobic, they only occupy anoxic ecological niches from which oxygenic organisms are excluded.

All extant anoxygenic photosynthetic bacteria contain only one type of reaction centre: either a type I RC resembling the cyanobacterial PSI; or alternatively, an anoxygenic type II RC that resembles the cyanobacterial PSII but lacks the capacity to split water (see Fig 1.2). These RC

Fig. 1.2 Comparison of oxygenic and anoxygenic photosynthesis pathways.

The electron transport pathways of photosynthetic of oxygenic and anoxygenic photosynthesis are remarkably similar. In oxygenic photosynthesis, as found in cyanobacteria, algae and plants, there are two photosystems called I and II. In anoxygenic photosynthesis, as found in some other bacteria, there is only one photosystem but this is very similar to either the type I or type II forms found in oxygenic photosynthesis. The vertical scale represents redox potential—the tendency to donate or accept an electron. Light generates the excited state, P*, of the primary electron donor, P. In type II reaction centres, the electron passes to a pair of membrane-bound quinones, Q_A and Q_B. In type I reaction centres, the electron passes to a chain of membrane-bound iron-sulphur centres, Fe-S, including water-soluble ferredoxin. Oxygenic photosynthesis couples photosystem II to photosystem I, allowing oxidation of water to generate electrons that eventually reduce the terminal electron acceptor, $NADP^+$. Anoxygenic photosynthesis has either a type I or a type II reaction centres, but never both. In all cases, a cytochrome complex oxidizes a quinone pool, Q, in a proton-translocating cycle coupled with ATP synthesis. Cyt cytochrome, PC plastocyanin.

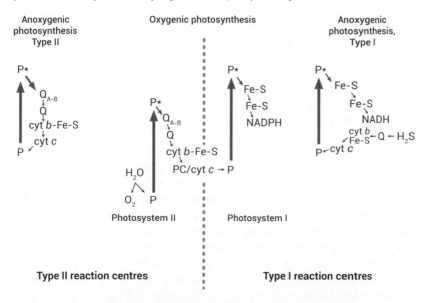

Scientific approach 1.2
Using phylogenetics and molecular clocks to study evolution

Phylogenetics, from the Greek *phylon* meaning tribe or clan, and *genetikós* meaning origin, is the scientific discipline of reconstructing and inferring the evolutionary relationships between organisms, genomes, genes, and proteins. By studying similarities between genomes, genes, or proteins, using powerful algorithms and statistical methods, it is possible to resolve how closely or distantly related two organisms or groups of organisms are to each other. This is done by producing phylogenetic trees, which also contain a wealth of evolutionary information about the nature of an ancestral state, the pace of evolutionary change, the time of evolutionary events, and the mechanisms of molecular evolution. These mechanisms include gene losses and duplications, and the exchange of genes between distantly related species, a process known as horizontal gene transfer (HGT).

As time goes by, a gene or protein will inevitably accumulate mutations. If one measures the speed at which a set of genes or proteins accumulates mutations in two distinct organisms, ie the rate of nucleotide (gene level) or amino acid (protein level) substitutions, it is possible to infer how much time has passed since the two organisms split into separate species. This branch of phylogenetics is known as molecular clock analysis. The technique is widely used across evolutionary biology to measure the timescales of rapid processes such as the evolution and spread of viruses that can occur over days or weeks. It can also be used to measure multi-billion-year processes, such as the evolution of organisms from LUCA to contemporary life forms including putative dates for the origin of major clades including eukaryotes, animals, and plants.

To calculate time using a phylogenetic tree, one needs to know in advance when a few evolutionary events occurred. These are known as calibration points and could be a bone or a tooth found in a cave, a reliably dated fossil, or the date of a key event such as GOE1, depending on the time scale of interest. Different organisms, different genes, and different parts of a gene evolve at different rates creating degrees of uncertainty in the measurement of time using molecular clocks. To account for this uncertainty, researchers use statistical analysis to quantify a range of probable ages for a particular evolutionary event of interest within a given phylogenetic tree.

Using phylogenetics it can be concluded that photosynthetic species are only present in 10 out of over 100 major clades of bacteria scattered through the Tree of Life, and only one group, the cyanobacteria, is capable of oxygenic photosynthesis. Through phylogenetics we can also determine definitively that the plastid genomes in photosynthetic eukaryotes originated from cyanobacteria and that some of the earliest evolving photosystems could split water. As noted in the main text, phylogenetic approaches are still being refined and in some cases there can be large margins of error, particularly for processes in the deep past. Nevertheless, they are powerful tools that add to other approaches such as fossil analysis and proxy geochemical methods.

types and their complex evolution are discussed in detail in Chapter 2. All the photochemical RCs are of a very ancient origin. However, the striking structural and functional similarities between type I and type II RCs indicate that photosynthesis originated only once, meaning that the evolution of all surviving RCs can be traced back to a single common source. The origin of photosynthesis required both the evolution of chlorophyll and bacteriochlorophyll pigments and the evolution of the protein machinery, the RC proteins, capable of binding these pigments onto a matrix that was able to harvest light and drive the extraction and transfer of electrons.

At present, it is not known exactly how the RCs originated because these protein complexes do not share obvious similarity with other known proteins. However, based on geological evidence it is thought that the process is most probably older than 3.4 Ga. Fossilized microbial structures, known as stromatolites, are found from this period and are thought to have been formed by the growth of photosynthetic organisms in shallow, well illuminated, coastal habitats. Over billions of years of evolution, photosynthesis diversified and became more complex, with each component in the process following a separate evolutionary pathway. In the next chapter, we will use the detailed insights gained from the structure and function of the photosystems during the past few decades, in combination with *phylogenetics* and new geological insights, to piece together the origins and early evolution of anoxygenic and oxygenic photosynthesis (see also Fig 2.1).

1.3   The origin of life

The Earth was formed about 4.6 Ga as a sphere of molten rock that gradually cooled until, possibly as early as 4.3 to 4.2 Ga, liquid water was able to accumulate (see Fig 1.3). By 3.95 Ga, there is indirect evidence of life from carbon isotope signatures. This means that the first living cells might have arisen within a relatively narrow time window of about 200 million years late in the Hadean Eon, about 4.1 or 4.2 Ga. By this time, there was plentiful water on the Earth, but it was still a highly volcanic planet with an atmosphere mainly consisting of nitrogen, carbon dioxide and water vapour, plus traces of methane and hydrogen. Because there was no free oxygen in the atmosphere or dissolved in the seas, it was only possible for anaerobic organisms to survive during this period.

Every living organism has two major prerequisites for successful evolution, namely (i) an efficient mechanism for reproduction and (ii) a barrier between its own unique genetic and metabolic machinery and the external environment.

Fig. 1.3 Timescale of the evolution of early life forms.

The Earth was formed about 4.6 Ga at the start of the Hadean Eon. By 4.3 to 4.2 Ga it has cooled suffi-ciently for the accumulation of large amounts of liquid water, although there was still considerable vol-canic activity and meteorite bombardments. Early life might have arisen before the start of the Archean Eon at 3.8 Ga. The Last Universal Common Ancestor of life or, LUCA, might date from this period and it gave rise to the two major Divisions of life, bacteria and archaea. The origins of oxygenic photosynthesis are still debated but could also date to the early Archean. However, any oxygen released by photosynthe-sis was rapidly taken up by minerals such as free iron. Eventually, rates of oxygen production exceeded its uptake by minerals and global levels rose from near zero to about 1% of present levels during GOE1 at about 2.3 Ga at the start of the Proterozoic Eon. Around this time, Earth became a mostly aerobic planet with a series of cell fusions and endosymbioses involving bacteria and archaea giving rise to eukaryotes containing mitochondrial and plastid organelles. During the Proterozoic, algae diversified to become the dominant photosynthetic producers and during GOE2 this resulted in another rise in oxygen levels to almost present-day values. After 0.5 Ga, the Phanerozoic Eon witnessed a fully oxygenated Earth with a protective ozone layer that facilitated the diversification of multicellular life and terrestrial colonization.

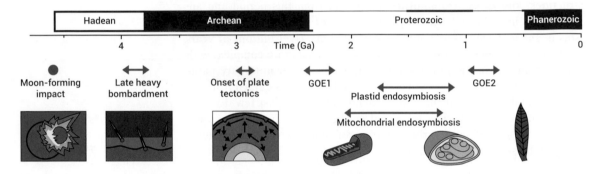

© 2018 H.C. Betts, M.N. Puttick, J.W. Clark, et al

Hence, all extant organisms on Earth are based on discrete membrane-bound cellular structures that reproduce via DNA/RNA-mediated genetic processes. It is generally (but not universally) believed that life probably originated in a so-called 'RNA world' in which very simple life forms used RNA, that functioned both as a genetic molecule able to copy itself and as RNA-based enzymes, called ribozymes, able to catalyse a limited range of biochemical reactions. The Darwinian selection of better-adapted versions of these early life forms would not have been possible unless each individual entity could be physically separated, both from other life forms and from the external environment. In other words, genetics-based evolution would not have been possible without some sort of a limiting membrane or other type of physical barrier between each replicating unit or cell.

This limiting membrane was provided by lipids. Many lipid molecules, such as fatty acids and their esters, can spontaneously form bilayer membranes and these could have encapsulated RNA-based genetic material and other complex organic molecules to form the first primitive cells. The physical location of such early life forms is unknown, but the two of the leading contenders suggested by researchers are (i) deep oceanic hydrothermal vents and (ii) so-called 'warm little ponds' (WLPs) on the surface of the Earth. The WLP hypothesis is a modern reframing of an idea originally proposed informally by Charles Darwin in the nineteenth century. This hypothesis has several attractive features including the occurrence of repeated wet and dry cycles that facilitate both polymerization of nucleotides to RNA and the spontaneous assembly of individual acyl lipid molecules into relatively stable macro-structures such as bilayer membrane-bound vesicles.

It is possible that the first free-living organisms to evolve in hydrothermal vents and/or WLPs were heterotrophic (see Fig 1.4). These cells were able to break down abiotically generated organic molecules into simpler products in order to generate energy for their metabolic processes. In this case, the energy was derived from the breakdown of abiotically generated chemicals that were already present in their environment. The energy released by such catabolic activities enabled these cells to synthesize their own complex molecules, such as carbohydrates, proteins and DNA, from simple inorganic precursors such as nitrates and CO_2. The organic molecules used by the earliest protocells were formed by geological or atmospheric processes, such as volcanic activity in undersea vents

Fig. 1.4 Flowchart showing autotrophic and heterotrophic lifestyles.

From the Start box, an organism is autotrophic if it obtains its carbon from elsewhere (Yes) or heterotrophic if it makes its own organic carbon (No). In turn, there are two types of autotroph – photoautotrophs and chemoautotrophs. Photoautotrophs, which include all photosynthetic organisms, use light as an energy source and can make their own carbon compounds. Chemoautotrophs, which include nitrogenfixing and iron-oxidizing bacteria, can fix CO_2 and derive energy from chemical processes. There are also two types of heterotroph – photoheterotrophs and chemoheterotrophs. Photoheterotrophs, which include several types of bacteria, can use light as an energy source but cannot fix CO_2. Chemoheterotrophs, which include the majority of heterotrophs including all animals and fungi, need to consume complex chemicals from other organisms to stay alive.

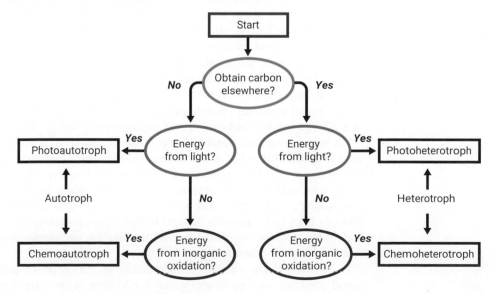

or the action of lightening on mixtures of organic molecules nearer the surface.

1.4   LUCA: the Last Universal Common Ancestor of extant life

The earliest discrete life forms were all unicellular (made up of single cells) and were prokaryotes (lacking a membrane-bound nucleus and complex internal organelles). It used to be assumed that the cells that evolved and diversified during the late Hadean and early Archean Eons were relatively simple compared with modern cell types. But it is now believed that these cells already contained much of the metabolic machinery found in modern organisms. For example, many of the early cells, whether autotrophs or heterotrophs, would have needed a wide range of metabolic pathways to synthesize their organic components. The period immediately after the dawn of cellular life was one of rapid evolution and experimentation, especially in terms of acquiring specific metabolic capacities required to exploit the available sources of nutrients in a given location. This period of metabolic experimentation probably lasted several hundred million years and many of the resulting cell lines died out without producing any descendants. However, one group of prokaryotes not only survived, it became the ancestor of all extant life on Earth. This small group of cells, or possibly just a single cell, used the same genetic code and amino acid chirality, and had the same core set of genes that are still present in all extant life forms. This cell(s) is commonly referred to as the Last Universal Common Ancestor of life or LUCA.

Origins of LUCA

LUCA probably arose in the early to mid-Archean Eon, about 3.9 to 3.4 Ga, although some molecular clock estimates date LUCA back to the late Hadean Eon, ie prior to 4.0 Ga. By this time, cellular life would have moved beyond its immediate region(s) of origin, whether these were in WLPs and/or deep hydrothermal vents, and had been able to colonize broader aquatic habitats. There is evidence that, by 3.2 Ga, the majority of the Earth's surface was covered by a single ocean, which would have meant that potential terrestrial habitats were limited during this period. A combination of genomic, biochemical, and phylogenetic evidence suggests that LUCA was a chemoautotroph with a H_2-requiring lifestyle. It has been previously assumed that LUCA was thermophilic (growing optimally at high temperatures, typically above 45°C). However, this has been challenged recently with evidence that LUCA was probably mesophilic (optimal growth between 20°C and 45°C). It has yet to be demonstrated whether LUCA had DNA- or RNA-based genetics. Several DNA-binding proteins can be traced back to LUCA so it probably contained DNA, but it is unclear whether LUCA could actually replicate DNA (see Fig 1.5).

LUCA was able to use carbon compounds such as formate, methanol, acetyl moieties, and pyruvate, which are known to be formed spontaneously from CO_2, and also had access to free metals and water under

Fig. 1.5 LUCA, the Last Universal Common Ancestor.

LUCA might have been a single cell or a population of cells and is the most recent life form from which all extant organisms are descended. In this model from 2020, LUCA is shown as the ancestor of two primary Domains of Life, bacteria (B) and archaea (A). The Asgard branch of the archaea gave rise to the precursor of all eukaryotes, which are descended from LECA (Last Eukaryotic Common Ancestor). However, it is also clear that LECA also had bacterial component derived from an endosymbiont as discussed in the main text.

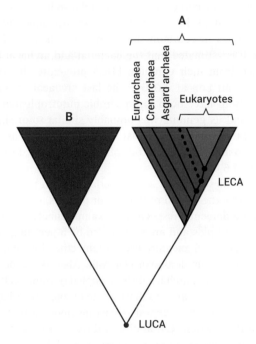

the conditions known to be prevalent at this time in Earth's history. LUCA used hydrogenase enzymes that were able to channel electrons from environmental H_2 to reduce ferredoxin. Ferredoxin is still the main source of high-energy reducing power in anaerobic organisms and is used in aerobic photosynthesis to generate NADPH. LUCA was therefore most likely a chemoautotroph able to convert the plentiful simple carbon compounds available in its environment into the more complex metabolites such as amino acids, lipids, and carbohydrates that are required for growth and reproduction. Recent evidence suggests that very small, but measurable, amounts of oxygen were present in some locations when LUCA evolved. Therefore, although LUCA was anaerobic, it probably already contained several enzymes, such as superoxide dismutase (SOD), that enabled it to withstand the presence of low levels of otherwise highly toxic reactive oxygen species (ROS) and free oxygen. This raises the interesting question as to whether small amounts of oxygen (either abiotic or biotic in origin) might have been present in the localized niches where LUCA evolved.

Origins of bacteria, archaea, and photosynthesis

It is likely that many diverse and novel groups of organisms evolved from the original LUCA over the next few million years, but only two of these groups are still extant, namely the two primary biological Domains, bacteria and archaea, that together make up all of life on Earth today (as discussed below, the eukaryotes are now increasingly recognized as a subgroup of the archaea). The earliest archaea were probably H_2-dependent methanogens, while the earliest bacteria would also have been obligate anaerobic chemoautotrophs. In both cases these strictly anaerobic prokaryotes had numerous genes in common that were most likely derived from the original LUCA. It is estimated that the bacterial and archaeal Domains had already diverged from their common LUCA progenitor by soon after the start of the Archean Eon at 4.0 Ga. The last archaeal common ancestor (LACA) was likely to have been an anaerobic autotroph that had evolved prior to 3.6 Ga and its genome was probably similar to or slightly smaller than those of extant archaea. The last bacterial common ancestor (LBCA) was probably also an anaerobic autotroph and is estimated to have evolved between 3.9 and 3.6 Ga.

Although the earliest archaea and bacteria are both unicellular prokaryotes, with no membrane-bound organelles or nucleus, they differ from each other in several fundamental respects. For example, bacterial cell membranes contain ester-linked lipids and are surrounded by a peptidoglycan cell wall. In contrast, archaeal cell membranes contain ether-linked lipids and are bounded by a pseudopeptidoglycan cell wall. Also, the gene transcription and translation systems of archaea differ completely from bacteria, although the former are strikingly similar to those of eukaryotes (see Chapter 3). Other differences include a methanogenesis-based metabolism that is only found in archaea, and their widespread ability to thrive in extreme environments, such as at very high temperatures, pressures, and salinity.

Following their divergence, possibly prior to 4.0 Ga, bacteria and archaea evolved along different trajectories, both of which are relevant to the origins and evolution of photosynthesis. Bacteria diversified into many different groups that included obligate and facultative anaerobes. Several bacterial groups eventually evolved various types of anoxygenic photosynthesis although the timing and mechanisms of this remain unclear. However, as noted above, only one distinctive group, the cyanobacteria, is capable of oxygenic photosynthesis. The bacterial origins of photosynthesis are explored in greater detail in Chapter 2. All extant photosynthetic mechanisms, whether oxygenic or anoxygenic, can be traced back to a single bacterial origin that occurred at least 3.4 Ga. Oxygenic photosynthesis resulted in a gradual change from a fully anoxic to a mainly oxic world. This in turn forced most (but by no means all) bacterial groups to become aerobic, including the α-proteobacterium-like organism that likely merged with an archaeal cell to create the first eukaryote at about 2 Ga (see section 1.5).

The archaea also quickly diversified from their common ancestor and, thanks to their more robust ether-based membrane lipids and heat-resistant enzymes, several groups were able to colonize novel habitats that were too extreme for most bacteria. Indeed, many extant groups of archaea are classified as extremophiles, meaning that they are adapted to living in very harsh environments that cannot support other organisms. Such environmental conditions include temperatures above the boiling points of water as found in marine hydrothermal vents (110°C to 121°C), very low, sub-zero temperatures in Antarctica (−14°C to −18°C), hyper-alkaline soda lakes (pH >11), hyper-acidic volcanic springs (pH <1), extreme pressures in deep oceanic trenches (>1,000 bar), and hyper-saline lakes (with salt concentrations of 320 g.l^{-1} or almost ten-fold that of seawater).

Research into the biochemistry and physiology of archaea has been hindered by the difficulty of cultivating them in the lab. This problem has been partially circumvented by the sequencing of increasing numbers of archaeal genomes in environmental and metagenomic sampling studies. The results are unexpected in showing that the archaea are by no means all extremophiles, but also include many species that are present throughout the full range of non-extreme, or mesophilic, habitats such as rocks, soils, oceans, and lakes. In addition to free-living species, many archaea are also present in various types of symbiotic associations with bacteria and eukaryotes. Indeed, archaea are now recognized as important components of the human gut microbiota alongside various bacteria, and eukaryotes. Some archaea even appear to have adopted a more predatory lifestyle with the capacity to ensnare bacterial prey. As discussed in the next section, emerging evidence suggests that one group of archaea related to the extant Asgard clade could have been the host cell that gave rise to the first eukaryotic organism (see Fig 1.5).

In summary, it is likely that photosynthesis initially developed in bacteria well before 3.0 Ga and possibly as early as 3.8 Ga. There is still some uncertainty about exactly how and when it evolved and in particular whether the oxygenic or anoxygenic versions came first (see Chapter 2). Whatever its origins, however, photosynthesis remained confined to bacteria for over a billion years. During this time, oxygenic photosynthesis in the cyanobacteria became the globally dominant photosynthetic mechanism. Thanks to the efficiency of oxygenic photosynthesis the cyanobacteria soon became the most important life forms in terms of their global primary production. Their metabolic activity resulted in the release of increasing amounts of oxygen into the atmosphere during the remainder of the Archean Eon, which lasted until 2.5 Ga.

1.5   Origins of eukaryotes

Eukaryotes include all the complex multicellular life forms, such as plants, animals, and fungi, that are such prominent features of today's biosphere as well as many ecologically important protists, such as the algal phytoplankton and the zooplankton. As discussed above, photosynthetic eukaryotes such as plants and algae dominate today's Earth in terms of their primary productivity and biomass. Although the evolutionary success of the eukaryotes has been considerable, it should not be overstated. The relative complexity of eukaryotic cells, with their extensive endomembrane systems and organelles, means that they are almost all restricted to relatively benign mesophilic habitats. Such habitats are characterized by a fairly narrow range of moderate environmental conditions with regard to their temperature, pressure, pH, and salinity. This is in contrast with the many extremophile archaeal and bacterial species discussed in the previous section that are able to flourish in much harsher environments. Happily for the eukaryotes, extensive areas of the Earth have been host to mesophilic habitats for the past 2 to 3 billion years, and this has enabled them to dominate most terrestrial and oceanic ecosystems in terms of their diversity and biomass. However, should environmental conditions eventually become more drastic, it is likely that eukaryotes would be unable to survive and the Earth would be dominated instead by extremophile archaeal and bacterial species.

Eukaryotes include relatively simple unicellular or colonial and larger, more complex, multicellular organisms. In all cases their cells contain a nucleus that houses the principal cellular genome, which is made up of several linear chromosomes stabilized by specialized proteins, such as histones. The nucleus is separated from the cytosol by a double nuclear envelope membrane that is contiguous with the endoplasmic reticulum membrane system. Eukaryotic cells also contain several types of membrane-bound organelles that can include mitochondria, lysosomes, Golgi, and plastids. It was previously thought that eukaryotes, alongside bacteria and archaea, made up the three Domains in what is called the 'Tree of Life'. Although the notion of three Domains is still much debated and is frequently presented in many biology textbooks, it has been increasingly challenged over the past decade. Multiple lines of genomic and phylogenetic evidence suggest that the ancestor of the eukaryotes arose from deep within the archaea where their closest relatives today are the Asgard group. This would mean that eukaryotes would be more correctly classified as a monophyletic group descended from within archaea rather than being regarded as a separate Domain of the Tree of Life.

Within the Asgard archaea, eukaryotes are most closely related to the Lokiarchaeota. For example, genes encoding several important classes of protein previously regarded as unique to eukaryotes, such as GTPases and SNAREs, have now been found in Asgard archaea genomes. Also, unlike many other archaea, Asgard species appear to be capable of facultative aerobic metabolism. The genomic composition of Asgard archaea suggests that complex eukaryotic features were already present before the acquisition of mitochondria. Such features including simple phagocytic capabilities (see below) in the putative archaeal host that gave rise to eukaryotes. However, eukaryotes also share many bacterial features, such as ester-based cell membrane lipids and mitochondrial oxidative respiration. It is evident therefore that eukaryotes have a combination of archaeal and bacterial features, suggesting that they might be descended from a composite cell derived from both ancestors. Therefore the original eukaryotic cell(s), known as the Last Eukaryotic Common Ancestor or LECA, would have been a chimeric organism derived from two (or possibly more) bacterial and archaeal progenitors (see Fig 1.5).

Until very recently, one problem with the above hypothesis was that neither archaea nor bacteria were thought to be capable of engulfing other cells via the process of phagocytosis, which was thought to be restricted to eukaryotes. This issue was partially clarified in 2020 with a report that a newly discovered Lokiarchaeota species had, for the first time, been cultured in the lab thus enabling its morphology and behaviour to be studied in detail. The genome of this species, called *Prometheoarchaeum*, contains many proteins that were hitherto regarded as unique to eukaryotes. However, it is the behaviour of this archaeon that is most interesting for our understanding of how eukaryotes evolved. Unlike all of the previously studied prokaryotes, its cells can produce long, branching protrusions possibly able to ensnare other organisms. This could allow them to capture, tether, and eventually engulf other cells, including bacteria. Another important feature is that this archaeon only grows together with various species of Methanobacteria in a form of symbiosis called syntrophy, where waste products from one partner are used as nutrients by another partner. Sometimes several microbial species form larger syntrophic consortia, as possibly occurred during the colonization of the terrestrial surface that eventually led to land plant evolution after 1.0 Ga (see Chapter 6).

A possible hypothesis for the origin of LECA is that an archaeal cell with similarities to *Prometheoarchaeum* initially formed a syntrophic relationship with an aerobic bacterial cell similar to extant α-Proteobacteria. Eventually the bacterial partner was engulfed by the archaeal cell and subsequently converted into a mitochondrial organelle via endosymbiosis. The resulting cells had genetic and morphological features derived from their archaeal host, plus an advanced capacity for oxidative respiration from their bacterial endosymbiont. These features enabled them to flourish in the increasingly oxygenated environment that was present during the Proterozoic Eon after 2.4 Ga. Because they inhabited mesophilic environments, it was more adaptive for the early eukaryotic cells to adopt ester lipids from their bacterial endosymbiont, which are more flexible under such conditions. These replaced their original archaeal ether lipids, which are now absent from eukaryotic membranes.

While this model of the origins of eukaryotes is currently favoured by some researchers, it is only one of several recent hypotheses for the origin of eukaryotes and readers should note that the topic remains a lively and fast-moving area of contemporary biology. The origin of eukaryotes and the role of endosymbiosis in the evolution of eukaryotic photosynthesis are discussed further in Chapter 3. The transfer of photosynthesis to eukaryotes was achieved by the uptake and sequestration of a cyanobacterial cell by an early heterotrophic eukaryote that was capable of capturing bacterial prey. The capture of the cyanobacterium was followed by its conversion into a dependant organelle called a plastid. This single event gave rise to all of the subsequent eukaryotic photosynthetic groups including algae and land plants as discussed in detail in Chapters 4 and 5.

1.6   Global impacts of photosynthesis: geology, atmosphere, and evolution

Photosynthesis is undoubtedly the most important biological process on our planet. As described in Chapter 4, during the past 3 to 4 billion years, one particular type of photosynthesis (ie oxygenic) has transformed the Earth in several fundamental respects.

This singular metabolic process has radically changed the composition of many minerals via oxidation; it has oxygenated the entire atmosphere and much of the oceans; it created the stratospheric ozone layer, hence reducing levels of harmful ultraviolet radiation; and the very oxygen that it generated in huge quantities went on to poison much of the anaerobic life on the Archean Earth. However, although oxygen was toxic to early life forms, it also provided an opportunity for the evolution of more complex aerobic multicellular organisms that were able to use oxygenic photosynthesis as a more efficient method of producing useful energy via aerobic respiration.

It is also the case that, were it not for photosynthesis, there would probably be few, if any, terrestrial (land-based) organisms. On such an anoxic, non-photosynthetic world, all aerobic life, including plants, animals, and fungi would be absent. What little life there was would be restricted to a narrow range of microscopic anaerobic prokaryotic extremophiles living close to a few scattered chemical and geological energy sources, such as volcanic vents in the deep oceans. In short, the entire biosphere has been radically altered and shaped by evolutionary changes that result directly from photosynthesis. For example, it is estimated that complex eukaryotic organisms, such as plants and animals, make up about 86% of our global biomass. Even more impressive is the estimate that 85% of this biomass is made up of photosynthetic organisms. This dramatically demonstrates the dominance of photosynthesis across the full range of global ecosystems.

An equally momentous side effect of increasing atmospheric oxygen concentrations due to photosynthesis was that some of the oxygen in the upper stratosphere was converted to ozone. This created a protective ozone layer that prevented much of the harmful solar UV radiation from penetrating as far as the Earth's surface. Prior to the establishment of a rudimentary ozone layer at about 2.3 Ga, levels of UV at the Earth's surface would have been sufficient to damage macromolecules such as DNA and some proteins, thereby precluding the emergence of terrestrial life. Once a basic stratospheric ozone layer was established some cyanobacteria were able to colonize some land surfaces by forming microbial mats, where they possibly formed syntrophic communities with other microbes. However, during this period, no multicellular eukaryotic organisms, with the possible exception of a few lichens, could live under the harsh conditions that were still present at the surface. It took over a billion years before a more effective ozone layer was formed, and before soils suitable for plant establishment arose due to a combination of physical weathering of rocks and the presence of microbial mats dominated by cyanobacteria.

Slow and partial oxygenation during the Archean Eon, 3.2 to 2.5 Ga

Oxygen gas is highly reactive and was virtually absent from the atmosphere of the early Earth. There are several abiotic processes that can generate very small amounts of oxygen gas. For example, water molecules can be split by high-energy radiation from sunlight or certain types of radioactive decay. However, these abiotic processes are much less efficient than the strictly biotic process of oxygenic photosynthesis. Therefore, if cyanobacteria, or similar oxygenic organisms, had not evolved, oxygen would have only been produced in such tiny quantities that it would have rapidly reacted with reductants, such as free metals, especially iron, and could not accumulate either in the oceans or in the atmosphere. However, there is geological evidence of localized so-called 'whiffs' of oxygen starting about 3.0 Ga, which is consistent with the activity of oxygenic biological organisms (see Fig 1.6).

The isolated oases in which cyanobacteria were active would have contained sufficient oxygen to be toxic to almost all other life forms in the immediate vicinity because these would have all been obligate anaerobic organisms, either archaea or other bacteria. The cyanobacteria themselves would have already developed mechanisms to detoxify the relatively large amounts of oxygen and ROS that they were producing as a byproduct of oxygenic photosynthesis. Neighbouring anaerobes, including other bacteria and archaea, would have experienced a strong selection pressure to develop similar mechanisms, such as antioxidants, to enable them to withstand this new form of abiotic (albeit biologically generated) stress. Therefore, during the Archean Eon, there was a gradual shift from a totally anaerobic biota to a more complex habitat with localized regions of facultative aerobic/anaerobic biota in an otherwise mainly anoxic environment. As the cyanobacteria continued to proliferate, the oxic oases became more prevalent until truly anoxic habitats became increasingly scarce. This meant that obligate anaerobic organisms became restricted to the few remaining oxygen-free niches, such as volcanic vents and deep ocean habitats.

At a planetary level, the Archean oxidation was very gradual, taking many millions of years. This slow change occurred because the Earth contained many reductive molecules such as hydrogen, methane and ferrous iron (Fe^{2+}) that would have removed the oxygen as quickly as it was produced. In the case of Fe^{2+}, its oxidation yields the solid compounds magnetite and haematite, which are insoluble in water. As oxygenation progressed, large deposits of magnetite and haematite accumulated at the bottom of the shallow oceans to form the so-called banded iron formations. These rock formations can still be in geological deposits and are evidence of oxygenation due to photosynthesis that continued throughout the Archean until all of the free iron was oxidized. Until the capacity of cyanobacteria to generate oxygen exceed the extent of such reductive sinks, oxygen could not accumulate, but after this point, levels of the free gas rapidly increased both in the air and water.

As cyanobacteria proliferated in both overall numbers and the type of habitats occupied around the world, their activity increased both the productivity and complexity of many microbial ecosystems. They created

Fig. 1.6 Changes in the composition of Earth's atmosphere since 4.0 Ga.

At the start of the Archean, about 4.0 Ga, the atmosphere was dominated by methane (orange), CO_2 (yellow) and nitrogen (blue). The amount of oxygen (green) was less than one millionth of current levels. From 4.0 to 2.4 Ga oxygen levels remained low although there are indications of localized, transient accumulations (whiffs), probably due mainly to cyanobacterial photosynthesis. Following GOE1 at about 2.4 Ga there was a sudden increase in oxygen levels to create a largely aerobic world, but at only about 1% of today's levels. After GOE2, at about 0.8 to 0.6 Ga, oxygen levels increased again to values similar to those of today. While GOE1 was mainly or entirely caused by cyanobacterial photosynthesis, GOE2 was due to a combination of algal and cyanobacterial photosynthesis.

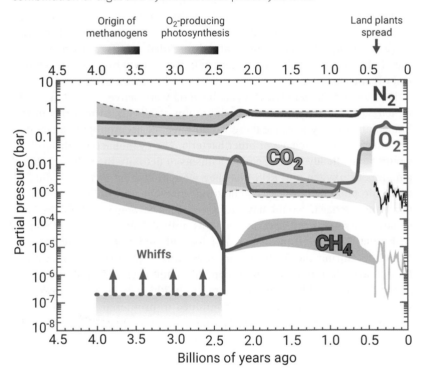

habitats, such as syntrophic microbial mats and stromatolites (see above) that were also home to a variety of heterotrophs plus other types of photosynthetic bacteria. Cyanobacteria were probably the first colonists of some terrestrial habitats, where they would have increased the weathering of rocks, and hence the flux of nutrients into the oceans. As the Archean Eon progressed, however, average global oxygen concentrations remained very low at below 0.1% of the present atmospheric level. Although almost all of the oxygen released by cyanobacteria during this period was rapidly removed by reductants, there was still scope for some localized accumulation in the oxic oases described above. In line with this, there is evidence of a significant increase in bacterial diversity between 2.85 and 3.33 Ga. Known as the Archean Expansion, this period saw the emergence in bacteria of numerous new gene families related to electron transport and respiratory pathways required for life in aerobic environments. This suggests that many new aerobic bacterial species were evolving, including the α-proteobacterium-like group that included the precursor of mitochondria in the future eukaryotes, that had yet to emerge at this point.

GOE1: the 1st Great Oxygenation Event in the early Proterozoic, ca 2.4 Ga

The boundary between the Archean and Proterozoic Eon occurred at 2.5 Ga. Shortly thereafter, between 2.45 and 2.32 Ga, significant quantities of oxygen began to accumulate in the air in an episode known as the 1st Great Oxygenation Event (GOE1). Before GOE1, atmospheric oxygen levels were as low as 10^{-5} of the present atmospheric level of ~21% by volume. By about 2 Ga, oxygen levels had risen to between 0.1 and 1% of the present amount. This turned out to be an irreversible event in that, thanks to oxygenic photosynthesis, the Earth's atmosphere has continued to contain significant, albeit variable, amounts of oxygen up to the present day. GOE1 decisively changed the Earth from a mainly anaerobic to a mainly aerobic environment. It also further accelerated the massive biological switch from anaerobes to aerobes as the principal lifeforms in terms of both biomass and biodiversity. Although GOE1 created an aerobic world, full oxygenation of the atmosphere and oceans to current levels of about 21% was not achieved until near the end of the Proterozoic Eon at about 0.85 Ga (see Fig 1.6).

The wider consequences of GOE1 for the evolution of photosynthesis, and particularly for its appearance in a new group of oxygenic eukaryotes, the plants and algae, are explored in more detail in Chapters 4 to 6. Thanks to the mitochondria acquired from their α-proteobacterium-like endosymbionts, eukaryotes were able to utilize the newly abundant oxygen for a more efficient form of energy metabolism based on oxidative respiration via glycolysis and the Krebs cycle. Together with the more efficient genetic machinery that was derived from their archaeal ancestors, the early eukaryotes were able to leverage their energetic advantages to develop into larger and more complex cells that became increasingly effective predators. Note that, at this stage, all eukaryotes would have been heterotrophs that relied on other organisms for nutrition.

Eventually, at about 2.0 Ga, one of these chimeric aerobic 'superorganisms' made the ultimate biological advance by ingesting, but not digesting, a cyanobacterium to create a photosynthetic eukaryote, the original ancestral alga. As described in Chapter 4, this appears to have been a unique event in biological evolution. Although cyanobacteria were, and still are, routinely ingested by eukaryotic cells, they are almost invariably either digested immediately or after a short period. The maintenance of the captured cyanobacterium and its eventual conversion into a dependent organelle was a protracted process that involved a great deal of genetic and metabolic adjustment by both the host cell and its photosynthetic organelle.

The result was the emergence of the first autotrophic eukaryotes. Thanks to their more efficient endomembrane systems and their capacity for aerobic respiration, photosynthetic eukaryotes were larger and faster growing than their prokaryotic ancestors and cousins, the cyanobacteria.

During the remainder of the Proterozoic Eon, photosynthetic eukaryotes gradually expanded their habitat range and diversified into a huge range of algal organisms as described in Chapter 5. By 1.8 Ga, the primordial alga had definitively converted its cyanobacterial endosymbiont into a dependent organelle. It had also diversified into the three main lineages that make up the Archaeplastida, namely red algae, green algae, and glaucophytes. By 1.6 Ga, some red algae made a giant evolutionary stride by becoming some of the earliest truly multicellular organisms, as they developed differentiated cell types. But the most successful group of photosynthetic eukaryotes was the green algae, or Viridiplantae, one subgroup of which evolved into the land plants. Part of the reason for the success of the green algae after 1.4 Ga might have been the increasing concentrations of nutrients in the oceans due to geological processes such as runoff caused by erosion or vulcanism. Green algae cells may have been better able to take advantage of these additional nutrients than cyanobacteria.

By 1.0 Ga, multicellular green algae, such as *Proterocladus*, are found in the fossil record and the photosynthetic eukaryotes begin to supplant cyanobacteria in terms of their overall contribution to global oxygen production. In addition to favouring the diversification of photosynthetic eukaryotes, the post-GOE1 Proterozoic Eon, from 1.8 to 1.0 Ga, witnessed the emergence of most of the major groups of heterotrophic eukaryotes. Sometimes collectively referred to as Protists, these were unicellular aerobic organisms that had a plentiful food source thanks to the abundance of bacteria, especially cyanobacteria, and algae, as well as other Protists. During the latter part of the Proterozoic all the major non-photosynthetic lineages emerged. These included the Opisthokonts, a group of flagellated eukaryotes that evolved into the multicellular ancestors of the fungi and metazoa (animals).

GOE2: the 2nd Great Oxygenation Event in the late Proterozoic, ca 0.8 Ga

Over 1.5 billion years after the GOE1, which occurred at about 2.3 to 2.4 Ga, a second, slower, but also highly significant major global oxygenation episode occurred between 0.8 to 0.6 Ga. This was the 2nd Great Oxygenation Event (GOE2), also known as the Neoproterozoic Oxygenation Event (NOE). GOE2 involved what turned out to be a momentous increase in atmospheric oxygen concentrations, from levels of about 1% to 10–15% (see Fig 1.6). On a global scale, the oceans had become fully oxygenated by 0.55 Ga and atmospheric oxygen concentrations were approaching modern levels. On the land, the supercontinent of Rodinia, which had been formed between 1.1 and 0.9 Ga, started to break up into smaller continental land masses after 0.8 Ga. By this time, algae were the major contributors to global photosynthesis, having overtaken the slower growing cyanobacteria as the most important primary producers after 1.0 Ga.

The onset of GOE2 coincided with the rise of multicellular green algae, such as *Cladophora*, which occurred at about 0.8 to 0.7 Ga. These algae, which were the ancestors of land plants, soon spread throughout the extensive oceans and shallow seas that covered nearly all of the Earth at this time. Between 0.6 and 0.4 Ga, there was an explosive growth of new forms of red, brown, and green algae. The importance of high oxygen levels for plant evolution may seem counterintuitive and it often been overlooked. Oxygen is a waste product of the photosynthetic process, but it is also the key substrate for mitochondrial aerobic respiration that is essential to growth and development. This is especially important in the case of the larger multicellular algae that gave rise to land plants. Plant processes such as reproduction are particularly energy demanding and are severely inhibited below 15% oxygen, ceasing entirely below 2.5% oxygen. This implies that the further evolution of algae into large and complex land plants was constrained by lack of oxygen until the atmosphere became fully oxygenated at the end of the Proterozoic about 0.55 Ga (Fig 1.6).

Another important factor is that, prior to 0.55 Ga, the amount of ionizing radiation at the Earth's surface might have been sufficient to preclude development of complex multicellular life. After 0.55 Ga, radiation levels were considerable attenuated by establishment of a more stable and concentrated stratospheric ozone layer. Another possible factor is the presence of a stronger dipolar magnetic field following the solidification of the inner core of the Earth. This deflects ionizing radiation from the solar wind and acts synergistically with the atmosphere to minimize the flux of UV or gamma radiation to the Earth's surface. The inner core of the Earth was probably not fully solidified until about 1.0 Ga so it was only after this time that it became possible for a few intrepid groups of organisms including some algal protists, cyanobacteria, and possibly some fungi, to begin the process of colonizing the land (see Chapter 6).

In addition to photosynthetic eukaryotes, many of the heterotrophic eukaryotic groups that had arisen after GOE1 underwent an additional diversification after GOE2. In all of these eukaryotic groups, whether photosynthetic and heterotrophic, the higher oxygen levels enabled them to increase their metabolic rates sufficiently to increase in both size and complexity. This was reflected in the appearance of many more types of multicellular organisms and the first metazoan fossils date from about 0.64 Ga. These marine eukaryotes soon diversified and were the ancestors of all extant animals. Similar to the argument for the evolution of complex plants being constrained by lack of oxygen, the same is probably even more true for animals and they too started their first diversification at the end of the Proterozoic about 0.55 Ga. While most early metazoans were soft-bodied, sessile filter feeders, by 0.55 Ga new motile forms capable of effective predation on similarly mobile prey had evolved.

The GOE2 therefore set the scene for the emergence of what is termed the 'Ediacaran Biota'. This group of organisms included all of the major taxonomic groupings of eukaryotes that are present in today's biota, such as the complex macroscopic multicellular algae and animals. The success of these new life forms was predicated on the creation of the highly oxygen-rich atmosphere and oceans that were able to sustain such large and active organisms. By the end of the Proterozoic Eon, increasing global temperatures enabled these multicellular algae and animals to diversify further in the so-called 'Cambrian explosion' that started about 0.54 Ga. This event saw the diversification of all of the extant groups of metazoans, although they were initially confined to aquatic habitats. At the same time, charophyte green algae were exploring freshwater habitats and developing adaptations to drying that eventually enabled them to colonize the land.

The evolution of some green algae into land plants, and the consequent establishment of terrestrial photosynthesis during the Phanerozoic Eon, after 0.55 Ga, is explored in more detail in Chapter 6. The land plants have developed novel photosynthetic traits in their methods for CO_2 concentration and fixation into carbohydrates. As discussed in Chapter 7, these traits, including C4 and CAM, are physiological adaptations by plants to environmental factors, such as water availability and temperature, that have appeared relatively recently (after 0.03 Ga) compared to photosynthesis itself. By studying photosynthesis in plants, algal and cyanobacteria, researchers hope to improve this vital process to help adapt to challenges such as global climate change.

Chapter summary

- Photosynthesis involves the use of solar energy to convert simple inorganic substrates into complex carbon-based compounds.
- Life probably originated about 4.0 Ga, as early cells used abiotically generated compounds as energy sources. Early life consisted of anaerobic prokaryotic bacteria and archaea in an almost completely anoxic world.
- Photosynthesis arose at least 3.4 Ga and maybe earlier. Ancestors of cyanobacteria developed oxygenic photosynthesis, creating localized oxic niches where aerobic organisms were able to evolve.
- Photosynthetic oxygen generated by cyanobacteria led to a global oxygen increase during the GOE1 about 2.4 Ga. This enabled the evolution of photosynthetic eukaryotes, the algae, which gradually diversified and eventually overtook cyanobacteria as the major oxygen producers.
- During the GOE2 after about 0.8 Ga, oxygen levels increased further and several groups of algae developed more complex multicellular forms and colonized new aquatic habitats.
- By about 0.55 Ga, some **charophyte** green algae became increasingly adapted to terrestrial conditions, evolving into all of the extant clades of land plants.

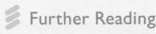

Further Reading

Catling DC, Zahnle (2020) The Archean atmosphere, *Science Advances* 6 (9), eaax1420. DOI:10.1126/sciadv.aax1420
Detailed review of the atmosphere on early Earth.

Deamer D (2019) Assembling Life: How Can Life Begin on Earth and Other Habitable Planets? Oxford University Press. ISBN: 9780190646387
Book examining possible life origins on Earth and other planets.

Doolittle WF (2020) Evolution: Two Domains of Life or Three? *Current Biology 30*, R159–R179. DOI:10.1016/j.cub.2020.01.010
Article from a noted authority discussing whether there are 2 or 3 Domains of Life.

Imachi H, Nobu MK, Nakahara N et al (2020) Isolation of an archaeon at the prokaryote–eukaryote interface. *Nature* 577, 519–525. DOI:10.1038/s41586-019-1916-6
Landmark article describing an archaeon with some eukaryotic characteristics.

Javaux E (2019) Challenges in evidencing the earliest traces of life. *Nature* 572, 451–460. DOI:10.1038//s41586-019-1436-4
Assessment of the evidence for very early life on Earth.

Taverne YJ, Caron A, Diamond C, Fournier G, Lyons TW (2020) Oxidative stress and the early coevolution of life and biospheric oxygen, In: *Oxidative Stress*, 67–85. DOI.10.1016/B978-0-12-818606-0.00005-5

Ślesak I, Kula M, Ślesak H, Miszalski Z, Strzałka K (2019) How to define obligatory anaerobiosis? An evolutionary view on the antioxidant response system and the early stages of the evolution of life on Earth, *Free Radical Biology and Medicine* 140, 61–73.
Three articles discussing when oxygenic photosynthesis might have evolved.

Service RF (2020) 'It will change everything': DeepMind's AI makes gigantic leap in solving protein structures, *Nature* 588, 203–204. DOI: https://doi.org/10.1038/d41586-020-03348-4
A new tool that could elucidate structures of photosynthetic membrane proteins.

Weiss MC, Preiner M, Xavier JC, Zimorski V, Martin WF (2018) The last universal common ancestor between ancient Earth chemistry and the onset of genetics. *PLoS Genetics* 14(8): e1007518. DOI: 10.1371/journal.pgen.1007518
Article discussing the possible nature of LUCA.

Williams TA, Cox CJ, Foster PG et al (2020) Phylogenomics provides robust support for a two-domains tree of life. *Nature Ecology and Evolution* 4, 138–147/ DOI:10.1038/s41559-019-1040-x
Article providing strong support for the '2 Domains of Life' hypothesis.

 Discussion questions

1.1 Describe the major impacts that photosynthesis has had on biological evolution.

1.2 What was LUCA and how does it relate to photosynthesis?

1.3 Discuss how eukaryotes evolved from prokaryotes.

1.4 How has photosynthesis changed the abiotic environment of Earth?

2 THE BACTERIAL ORIGINS OF PHOTOSYNTHESIS

Learning objectives

- Critically discussing different hypothesis for the evolution of photosynthesis
- Describing the evolution of type II and type I reaction centres
- Enumerating a minimum set of requirements for the origin of oxygenic photosynthesis
- Giving an overview of the evolution of chlorophyll synthesis

2.1 Introduction

In Chapter 1, we briefly surveyed the momentous roles of photosynthesis in the evolutionary history of the Earth. This chapter is concerned with a more detailed investigation of *where*, *when*, and *how* photosynthesis originated and then evolved in non-eukaryotic organisms. The broad answer to the 'where' question is 'in bacteria'; 'when' is still unresolved but was probably in the early-to-mid Archean Eon 'between 4 and 3 Ga' (billion years ago); and 'how' is currently the subject of considerable controversy as we will see below. Some of the best accepted geological evidence for the earliest photosynthesis comes from marine sedimentary deposits in rocks from the Buck Reef Chert in South Africa dated to 3.4 Ga. These deposits have unambiguous traces of life and their geochemical context is consistent with anoxygenic photosynthesis powered by H_2. Other even older rocks dating from about 3.8 Ga that could harbour geochemical signatures consistent with photosynthesis have been found in the Isua Greenstone Belt in Greenland. Therefore it is possible that anoxygenic photosynthesis had already evolved well before 3 Ga, at a time when the Earth was still a highly anaerobic planet (see Chapter 1).

2.2   Origin of photosynthesis

The timing of the origin of oxygenic photosynthesis is still controversial, but it had already appeared by about 2.4 Ga, which is when a steep increase in O_2 concentration can be detected in the geological record.

At this time, O_2 concentrations suddenly rose by several orders of magnitude from below 10^{-6} of the present atmospheric level (PAL) to about 10^{-2} PAL (see Fig 1.6). As outlined in Chapter 1, this process is known as the 1st Great Oxygenation Event (GOE1). However, multiple independent lines of evidence indicate that the origin of oxygenic photosynthesis antedated this event by hundreds of millions of years and bursts or 'whiffs' of O_2, and localized O_2 oases are thought to have occurred during this time. Evidence from redox proxies for O_2 in ancient rocks suggest that oxygenic photosynthesis had already evolved by 3.0 Ga. For example, well-preserved fossilized microbial mats resembling extant cyanobacteria are present in rocks dating from 3.2 Ga in the Barberton Greenstone Belt in South Africa. Further evidence for oxygenic photosynthesis has been reported, albeit not universally accepted, from rocks more than 3.7 Ga in Greenland. Unfortunately, rocks older than 2.5 Ga make up less than 5% of those that still survive, so the older samples are not necessarily representative of the geology of the period.

Prior to the **primary endosymbiosis** event at about 2.2 Ga (see Chapter 4), photosynthesis based on chlorophylls and bacteriochlorophylls was only present in the bacteria. Therefore, the origin of life and the early division of the two major domains of prokaryotes, archaea and bacteria, must have occurred prior to the origin of photosynthesis. If there was photosynthesis at 3.4 Ga or earlier, it follows, then, that the division between archaea and bacteria had already occurred by then. A few authors have raised the possibility that the origin of life itself required light and that LUCA was therefore photosynthetic, possibly using energy from abiogenic photochemical reactions, although this is a minority view. Assuming, as most researchers do, that photosynthesis originated in bacteria, it has still been difficult to determine at what point it occurred in evolution. This is because, in the absence of a definitive fossil record, it is challenging to determine what forms of life inhabited the planet before 3.4 Ga.

Some important questions include the following:

- Had any extant lineages of bacteria appeared by 3.4 Ga?
- Was the most recent common ancestor (MRCA) of bacteria photosynthetic?
- Did photosynthesis originate in an ancestral lineage that predated the diversification of the major extant groups of bacteria?
- Or, alternatively, did photosynthesis originate in one of the extant groups of photosynthetic bacteria?

These are questions that have been debated for several decades and have still not been answered conclusively. One important point is that, thanks to metagenomics and environmental sampling, many new and radically different types of bacteria and archaea have been discovered during the past few years. These discoveries have completely changed our perspectives on the evolution of photosynthesis and the nature of the entire Tree of Life.

There are at least two key evolutionary innovations required for the evolution of photosynthesis that will be discussed further here. First is the evolution of the reaction centre (RC) proteins. RC proteins bind chlorophylls and other cofactors to form fully assembled reaction centres. These RCs, also known as photosystems, carry out the primary photochemical reactions of photosynthesis. Second is a requirement for the evolution of biosynthetic pathways of chlorophylls and related pigments. Recent discoveries relating to both of these innovations are leading to new ideas and hypotheses about how, where and when photosynthesis evolved on Earth.

2.3   Photochemical reaction centres

Photosystems are pigment-protein complexes that harvest light energy when a chlorophyll molecule is excited by a photon. The excitation energy absorbed by a light-harvesting chlorophyll is transferred within few nanoseconds to a set of specialized core RC chlorophylls.

Upon excitation, these RC chlorophylls initiate a chain of oxidation-reduction reactions resulting in a flow of electrons in a process known as charge separation (Fig 2.1). The first specialized chlorophyll to initiate electron transfer, thereby becoming oxidized, is called the primary electron donor. The released electron is received by an adjacent chlorophyll called the primary electron acceptor. This photochemical reaction generates a highly reducing anion radical and a highly oxidizing cation radical. A series of rapid electron transfer reactions occurs, out of the anion radical and into the cation radical. Each of these reactions involves an increase in the distance between the charged radicals resulting in the formation of a series of radical pairs that are successively more stable.

The photosystems act as molecular machines that transform light energy into chemical energy and can be regarded as Nature's equivalent of photovoltaic cells. The flow of electrons triggered by the charge separation process is used by the cell to power all of its metabolic reactions, including the fixation of atmospheric CO_2 to produce sugars. One of the reasons it is suspected that photosynthesis originated only once is that all photosystems share a common structural architecture and functional core (Fig 2.2), namely a dimeric pigment-protein complex. Such photosystems are made up of two evolutionarily related monomeric RC protein subunits. A photosystem is homodimeric if both subunits are encoded by a single gene and the two monomers are identical. Alternatively, a photosystem is heterodimeric if each monomer is encoded by a different but related gene, giving rise to two subunits with different amino acid sequences.

Fig. 2.1 Charge separation during photosynthesis.

(a) Here, P represents the primary electron donor, A1 and A2 represent primary and secondary electron acceptors respectively, and A3 a terminal electron acceptor. D1 and D2 denote consecutive electron donors. Thus, in a photosystem the excitation of the main photochemical pigment P results in electron transfer to A1, transferring its negative charge (A1⁻), and retaining a positive electron hole (P⁺). In a second electron transfer event, the donor returns P⁺ to its reduced state, and A1⁻ pass its extra electron on to A2, increasing the distance between the positive and the negative charge. **(b)** This scheme shows a highly simplified schematic configuration of the photosystem redox cofactors (dark circles) named as in panel A on the left, and as in PSII on the right. In PSII the charge separation is slightly more complicated than shown in **a** because the terminal acceptor Q_B can accept two electrons, the Mn_4CaO_5 cluster can accumulate four positive charges, each step occurs at a different rate, and the redox chlorophylls (P_{D1}, Chl_{D1}, P_{D2} and Chl_{D2}) behave both as a single molecule and as individual molecules. In PSII the ultimate electron donor is water, which is only oxidized after four charge separation events have occurred in a catalytic cycle.

(a)

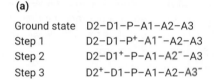

Ground state	D2−D1−P−A1−A2−A3
Step 1	D2−D1−P⁺−A1⁻−A2−A3
Step 2	D2−D1⁺−P−A1−A2⁻−A3
Step 3	D2⁺−D1−P−A1−A2−A3⁻

(b)

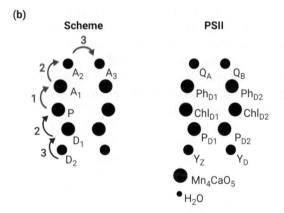

    Thanks to high-resolution X-ray crystallographic studies, we now know the details of RC protein structure down to the level of individual atoms in several bacterial species. An RC protein has two domains, an antenna domain and a core domain (Fig 2.2). The antenna domain binds the pigments used to harvest light. The harvested light is then channelled to the core part of the protein where the redox pigments are bound. Each RC monomer consists of a trans-bilayer protein with 11 transmembrane α-helices. The chlorophylls of the antenna involved in light harvesting are bound by the first 8 helices, and the core redox pigments are bound by the last 3 helices. The number of antenna chlorophylls varies from 31 to 90 depending on

Fig. 2.2 Reaction centre protein architecture.

(a) Monomeric secondary structures of RC proteins. PSI and PSII are photosystem I and II respectively, HbRC denotes the type I RC of Heliobacteria, and PbRC denotes the type II RC of Proteobacteria (Purple Bacteria). Vertical rectangles represent transmembrane α-helices, horizontal and tilted ones represent α-helical folds. Classically the antenna domain has been assigned to the first six transmembrane helices as highlighted in violet, and the core domain to the last five helices are shown in orange. However, light-harvesting pigments are found up the 8th helix, while only the last three helices (9th to 11th) directly bind the redox chlorophylls.
(b) This shows a view from above of how the transmembrane helices are positioned relative to each other. Notice that helix 11th is positioned in between 9th and 10th. This brings 11 closer to 6 in the antenna, as these two interact on the luminal side of PSII and HbRC.

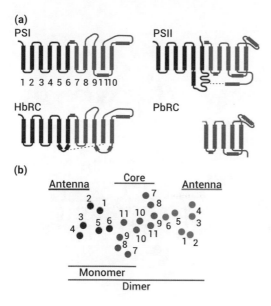

the photosystem and species, while the core redox pigments are made up of three pairs of symmetrically bound chlorophylls in a highly conserved orientation (see Chapter 3 for more details of RC structure and function). Despite the variation in the number of chlorophylls present in the antenna complex of different photosynthetic organisms, many of these chlorophylls occupy conserved positions through all known RC proteins in the Tree of Life.

Photosystems come in two forms: type I and type II reaction centres (Fig 2.3). Type I RCs contain an Fe_4S_4 cluster (F_X), which connects each monomer and works as the final electron acceptor bound by the RC proteins. Type I RC proteins have 11 transmembrane α-helices and are present as homodimers in anoxygenic species and heterodimers in oxygenic species. In oxygenic photosynthesis the heterodimeric type I RC is known as Photosystem I, or PSI. In contrast, type II RCs lack Fe_4S_4 clusters and instead the two monomers are connected by a single Fe atom (non-heme iron) and all are heterodimers. In type II RCs the terminal electron acceptor is a quinone, a lipid-soluble cofactor involved in redox transfer reactions.

Fig. 2.3 High-resolution structures of type I and II reaction centres (RCs).

(a) Anoxygenic Type II reaction centre of tpurple bacteria. Only the RC core subunit L is shown surrounded by the light harvesting complex LH1 (purple ribbons). **(b)** Oxygenic cyanobacterial PSII. The core of PSII is comprised of the reaction centre subunits D1 and D2, and the core antenna subunits CP43 and CP47. **(c)** The homodimeric Type I RC of the Heliobacteria with a single subunit known as PshA at its core. **(d)** Cyanobacterial heterodimeric PSI with a core of two subunits known as PsaA and PsaB. The different redox cofactors at homologous positions between different photosystems are shows as: P (photochemical pigment), M (monomeric chlorophylls), A (primary acceptor), and Q (quinone). Z denotes a corebound antenna chlorophyll retained in PSII and Type I RCs, but lost in the anoxygenic Type II RCs where a new light harvesting system also evolved. Monomers displaying the main redox cofactors are in the left column and full dimeric configurations are in the right column. Transparent grey ribbons mark the core domain and orange ribbons the antenna domain. Cofactors involved in charge separation are shown as orange ticks, antenna (bacterio)chlorophylls are shown in green lines, with the exception of that marked as Z; carotenoids are shown in red lines. Type II RCs contain a nonheme Fe^{2+}, while Type I RCs contain an iron-sulphur cluster, Fe_4S_4.

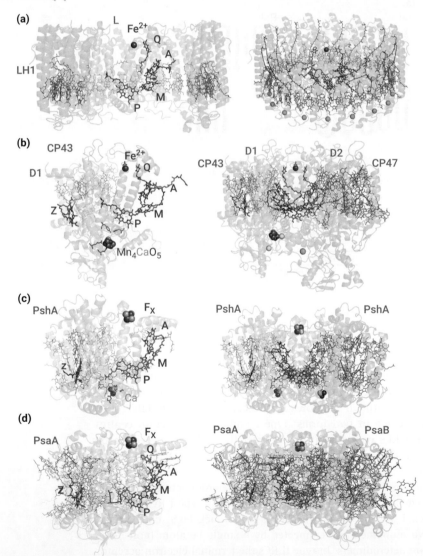

Case study 2.1
Role of quinones in reaction centres

In photosynthetic reaction centres, quinone molecules function as electron transfer cofactors. In type II RCs there are two quinones that occupy symmetric positions, known as Q_A and Q_B. Q_A is tightly bound to the protein complex and it is a one-electron shuttle. Its main role is to reduce Q_B. In contrast Q_B is highly mobile and once it has received two electrons and two protons, it dissociates from the binding site and is exchanged for an oxidized quinone (See Chapter 3 for more details). In type I RCs the roles of quinones are much less clear and have been debated for decades. In heterodimeric PSI, two quinones are bound at symmetric positions between the chlorophyll electron acceptor and F_X (Fig 2.3). These are one-electron shuttles and are not exchanged with the pool of quinones that reside in the photosynthetic membranes as in type II RCs. However, their position in the RC core is similar to that found in type II RCs. In contrast, in homodimeric type I RCs the quinones are loosely bound, or not at all, and are not required for electron transfer to F_X.

More recently, new evidence has suggested that some homodimeric type I RCs might have a dual function. Thus, under low-light conditions the RC reduces F_X, but under high light the cells enter an over-reduced state, meaning that the electrons cannot leave the RC. In this 'jammed' state, the RC might instead divert electrons to quinones, resulting in their complete reduction, and thereby operating in a way that resembles type II RCs. Nevertheless, clear evidence of binding sites for quinones within the structure of the homodimeric RCs has not been forthcoming. Based on these experimental observations and other rationales, it has been suggested that the earliest RC may not have had an F_X and instead resembled a type II RC at its acceptor side.

Quinones have several roles in RCs as described in Case study 2.1. Type II RCs are also present in two versions, the oxygenic type II RC and the anoxygenic type II RC. The oxygenic type II RC is also known as photosystem II, or PSII. In PSII, an RC monomer is split into two distinct subunits. One subunit is made up of the first six helices and the other subunit is made up of the last five helices. The heart of PSII is made up of four proteins: the two RC core subunits, D1 and D2, and the two antenna subunits, CP43 and CP47. In addition, to the core subunits a PSII complex contains several accessory proteins with various functions and which are required to optimize and regulate function (Fig 2.4).

PSII is unique among all RCs due to the presence of the water-oxidizing complex, also known as the oxygen-evolving complex. The water-oxidizing complex is an inorganic Mn_4CaO_5 cluster that it is bound

Fig. 2.4 Subunit composition of various photosynthetic RCs.

Each type of RC is associated with a unique series of ancillary subunits. The simplest is the homodimeric type I RC of Heliobacteria, made of two subunits, PshA and PshX each with one transmembrane helix. In the Chlorobi, three additional subunits are found surrounding F_X, including a light harvesting complex, FMO, and an additional ferredoxin-like subunit, PscB with two additional iron-sulphur clusters. This also includes PscC which is the electron donor to the core, a heme containing protein with three transmembrane helices that is tightly bound to the RC. Both PSI and PSII and characterized by a larger number of peripheral subunits with various functions. Like the Chlorobi type, PSII also has a ferredoxin-like protein attached close to F_X. PSII has several additional peripheral subunits and extrinsic polypeptides that protect the Mn_4CaO_5 cluster. Anoxygenic type II RCs are surrounded by LH1 and are additionally characterized by the presence of a tetraheme cytochrome that serves as an electron donor to the photochemical pigments.

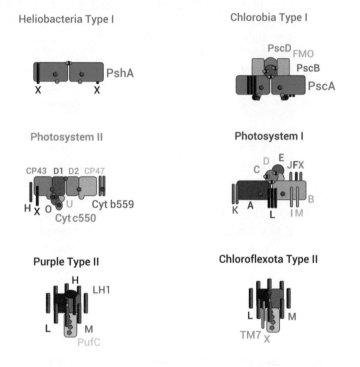

(coordinated) by the D1 and CP43 proteins (Fig 2.3B and Fig 2.5). This is the catalytic site where two water molecules are split into $4H^+$ and $4e^-$, with the release of O_2 as a by-product. The mechanism of water oxidation will be discussed in Chapter 3. In order to split water, PSII has evolved the capacity to generate 1.2 V of oxidizing power upon charge separation. In contrast, all other RCs only generate between 0.2 and 0.5 V of power. The oxidized chlorophyll cation then oxidizes a tyrosine residue bound by D1, Y_Z, which in turn oxidizes the Mn_4CaO_5 cluster (Fig 2.5). After four charge separation events, the cluster accumulates four positive charges, which are coupled to consecutive water deprotonation reactions. In the final step, an O-O bond is formed, and the cluster regenerates its most

Fig. 2.5 Structure of the PSII water-splitting complex at atomic resolution.

Water splitting is the key reaction of oxygenic photosynthesis and these diagrams show the Mn_4CaO_5 cluster that is responsible for this unique chemistry. **(a)** Location of the Mn_4CaO_5 cluster within a PSII complex. The protein scaffold of the complex is shown in transparent grey ribbons. The core subunit that binds the cluster, D1, is shown in orange, with the cluster shown as spheres within D1. Chlorophyll molecules are shown in green sticks. **(b)** Atomic detail of the cluster, purple spheres represent Mn atoms, red spheres represent oxygen atoms, and the green sphere is the Ca. The cluster is bound by residues from D1 (grey sticks) and the CP43 subunit (orange). Green dotted lines denote hydrogen-bonds, while black dotted lines coordination bonds.

(a)

(b)

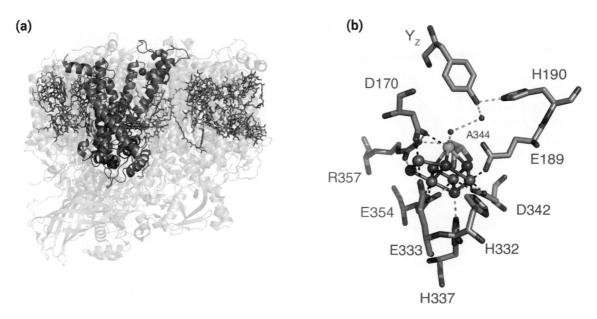

reduced state by taking four electrons from the two water molecules in a single step. The anoxygenic type II RCs lack a Mn_4CaO_5 cluster and therefore cannot split water, so their activity does not result in O_2 evolution. Anoxygenic type II RCs lack CP43 and CP47, and consist only of a core domain made up of two homologous subunits known as L and M. They have also evolved a novel antenna system, LH1, in a ring structure around the RC core (Fig 2.3).

Currently there are eight bacterial phyla with species known to have photosystems (see Table 2.1), although there is some evidence from metagenomic studies that more **phototrophic** clades might exist. The only bacterial group capable of oxygenic photosynthesis is the cyanobacteria. All photosynthetic eukaryotes (algae and land plants) acquired oxygenic photosynthesis following their incorporation of an endosymbiotic cyanobacterium that was transformed into a chloroplast organelle (see Chapter 4). All other groups of bacteria with RCs are anoxygenic. Oxygenic photosynthesis is the only photosynthetic process to use both types of RC in series as follows. Following light-driven water oxidation on PSII, reduced quinone molecules diffuse through the thylakoid membrane to the cytochrome b_6f complex, where they become oxidized with the resulting electrons being shuttled to PSI via a soluble one-electron carrier such as

Table 2.1 Clades of phototrophic bacteria

Phylum	RC	Core subunits	Oligomer status	Chlorophyll type	Discovery	Trophic mode
Cyanobacteria	I	PsaA, PsaB	Heterodimer	Chl *a, b, d, f*	Early 19th century	Photo-autotroph
	II	D1, D2 and CP43, CP47	Heterodimer			Mixotroph
Proteobacteria (*Purple bacteria*)	II	L, M	Heterodimer	Bchl *a, b*	Late 19th century	Photo-autotroph
						Mixotroph
						Photo-heterotroph
Chloroflexi (*Green non-sulphur bacteria, filamentous anoxygenic phototrophs*)	II	L, M	Heterodimer	Bchl *a, c*	1974	Photo-autotroph
	I	PscA	Homodimer		2020	
Eremiobacterota	II	L, M	Heterodimer	--	2018	Potentially photo-autotroph
Gemmatimonadetes	II	L, M	Heterodimer	Bchl *a*	2014	Photo-heterotroph
Firmicutes (*Heliobacteria*)	I	PshA	Homodimer	Bchl *g*	1983	Photo-heterotroph
Chlorobi (*Green sulphur bacteria*)	I	PscA	Homodimer	Bchl *a, c, d, e*	1912	Photo-autotroph
Acidobacteria	I	PscA	Homodimer	Bchl *a, c*	2007	Photo-heterotroph

plastocyanin or a free cytochrome. PSI then powers up the electrons in a second light-driven reaction to generate a strong reductant, ferredoxin, which is able to drive such essential photosynthetic functions as CO_2 or N_2 fixation.

It is important to note that there are no anoxygenic photosynthetic bacteria described to date that encode in their genomes both a type II and a type I RC. Therefore, a common trait in anoxygenic photosynthesis is that only one RC is used, either a type I or a type II but never both. Another major difference between oxygenic and anoxygenic photosynthesis is that the former exclusively uses chlorophyll as the main photochemical pigment. In contrast, the main pigment in anoxygenic photosynthesis is bacteriochlorophyll, as shown in Fig 2.6. In metabolic terms, bacteriochlorophylls are made from a precursor of chlorophyll *a*. The simplest bacteriochlorophyll known is bacteriochlorophyll *g*, which is an isomer of

Fig. 2.6 **Structural comparison of chlorophylls and bacteriochlorophylls.**

The main distinguishing feature between chlorophylls and bacteriochlorophylls is that in the latter, one of the double bonds in the pyrrole unit (red arrows) is reduced to a single bond. This shifts the peak of main absorption from the red into the far-red and infra-red. Other modified chlorophylls and bacteriochlorophylls exist where the rings have been decorated with additional modifications to fine-tune the absorption spectrum according to the organism's requirements, as shown in the red square for bacteriochlorophyll *a*.

chlorophyll *a*   farnesyl bacteriochlorophyll *g*   phytyl bacteriochlorophyll *a*

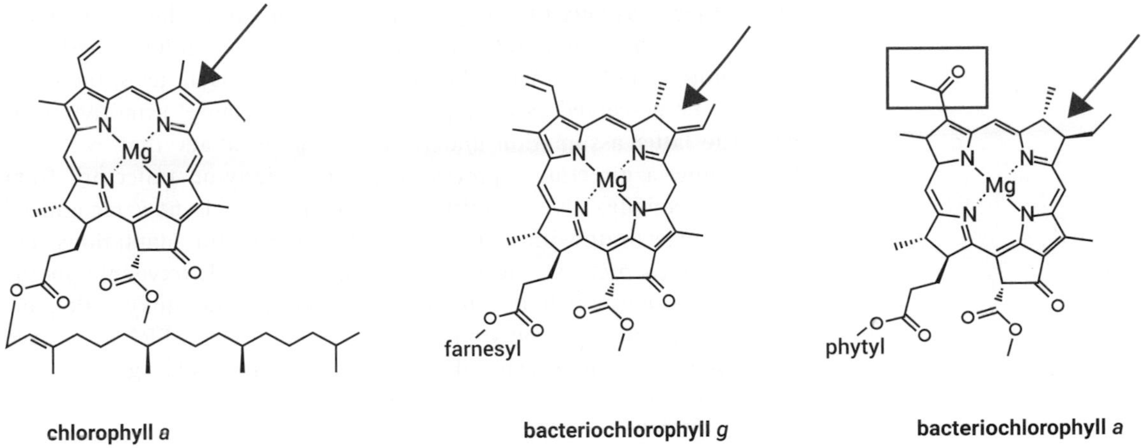

chlorophyll *a*. Bacteriochlorophyll *g* is exclusively used by Heliobacteria, the only photoheterotrophic group of the Firmicutes. All other anoxygenic phototrophs use bacteriochlorophyll *a*, which requires three additional enzymatic steps for its synthesis from the chlorophyll *a* precursor. All homodimeric type I RCs use a mixture of bacteriochlorophylls and chlorophyll *a* as their main photochemical pigments involved in charge separation.

2.4   Evolution of reaction centre (RC) proteins
Distribution of RC proteins

> To date, over 100 bacterial phyla have been defined taxonomically, many of which have never been cultured and are only known from environmental metagenome projects. Strikingly, only eight of these phyla have photosynthetic members (see Table 2.1).

Therefore, despite its apparent advantages in terms of food acquisition, photosynthesis is a relatively rare trait within the known diversity of extant prokaryotes. The capacity to make photochemical RCs is scattered across bacteria and all known photosynthetic species have close relatives that are non-photosynthetic. Given that photosynthesis originated at least 3.4 Ga, the scattered distribution of photosynthesis might be explained by factors such as gene losses, horizontal gene transfer (HGT) events and/or extinction of photosynthetic clades. The importance of gene losses relative to HGT events has been a point of contention in studies of the evolution of photosynthesis. But there is now unambiguous evidence that both processes have contributed substantially to today's diversity of photosynthetic bacteria. Interestingly, it is now known that photosynthesis has also been lost on many occasions in eukaryotes, including both algae (see Chapter 5) and plants (see Chapter 6).

Until recently it was assumed that anoxygenic photosynthesis evolved before oxygenic photosynthesis. There are two main reasons for this. Firstly, there was very little oxygen on the Earth during the first two billion years of its history. Secondly, because oxygenic photosynthesis is a much more complex process than anoxygenic photosynthesis, the latter tends to be perceived as a simpler and therefore more 'primitive' process. The latter assumption underlies a common fallacy that evolution is somehow a directional process with an underlying trajectory from simple to complex living entities. This assumption is now recognized as being inaccurate with many examples of successful adaptations that involve both structural and genomic simplification. If oxygenic photosynthesis, which uses both types of RC, emerged from anoxygenic photosynthesis, which uses only one type of RC, both types of RC must have come together in an ancestor of the cyanobacteria. This has given rise to the first two hypotheses for the evolution of photosynthesis, as shown in Fig 2.7A & B.

In the first hypothesis, it is proposed that the two RC types originated from an ancestral gene duplication event. If that is the case, then the ancestral photosynthetic organism evolved two reaction centres. This ancestral organism is assumed to be anoxygenic, and eventually gave rise to the oxygenic cyanobacteria. The ancestral two-reaction centre anoxygenic phototroph is colloquially referred to as the proto-cyanobacterium. After this duplication, the scattered distribution of RC in other bacteria is explained by multiple losses of RC genes across the Tree of Life, or alternatively by HGT before the evolution of photochemical water oxidation by PSII. The second hypothesis proposes that RCs diversified into two distinct types, not from a gene duplication event, but as two lineages of bacteria became adapted to different environments and metabolic needs. In this case, the scattered distribution represents HGT events from other anoxygenic phototrophs that enabled cyanobacteria to obtain their two RCs. Once the two RCs were present within the same membrane in a cyanobacterial ancestor, water oxidation could arise in PSII.

Although these two hypotheses appear mutually exclusive, they share a common feature. Both propose a transitional stage before the evolution of oxygenic photosynthesis that requires two anoxygenic RCs, a type II and a type I, working in series. Although it is usually taken as a fact that anoxygenic led to oxygenic photosynthesis, no evidence has yet been provided in support of this hypothetical transition. Thanks to more recent molecular studies, we now know the function and structure of the photosystems in much greater detail. This information, in combination with data from molecular phylogenetics, has given us a much more nuanced view of early RC evolution, which has led to these classic hypotheses being challenged. Much of this evidence has come from the lab of

Fig. 2.7 Three theories about the evolution of photosynthesis.

(a) This hypothesis postulates that the earliest RC started as type I (I). After a gene duplication event type II RC (II) then evolved from type I. Thus, the ancestral bacterium capable of photosynthesis had both types but did anoxygenic photosynthesis. The scattered pattern of photosynthesis across the tree of bacteria is due to loss of one type, loss of photosynthesis, or HGT (red arrows). The ancestor with the two anoxygenic RCs (red box) then evolved into cyanobacteria after developing water oxidation chemistry (green box). **(b)** A second scenario proposes that instead of a duplication, the two RC types evolved from individual organisms diversifying into new species. Here, oxygenic photosynthesis originated after both types were acquired via HGT. In both **(a)** and **(b)**, the nature of the ancestral RC could be of either type, or even a hybrid RC mixing traits of both. **(c)** Similar to A, but instead proposes that the divergence of type I and II reaction centres was linked to the origin of water oxidation.

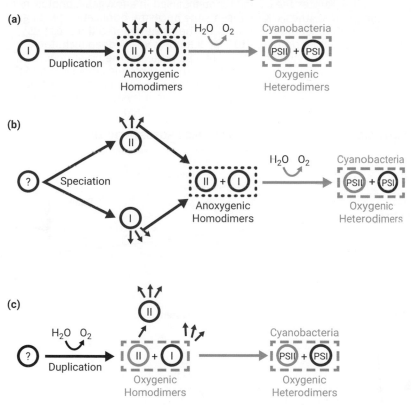

one of the authors of this book (TC) and is supported by the other author (DM). This has led to a third hypothesis that involves an alternative evolutionary scenario for photosynthesis as shown in Fig 2.7C and described in more detail in the following sections. It is expected that further research will eventually shed light on which, if any, of these three hypotheses is more accurate, or indeed if there is a completely different explanation of how photosynthesis evolved.

Evolution of type II RCs

Each reaction centre type is monophyletic. That means that the individual type II RC proteins, D1, D2, L and M, share a most recent common ancestor to the exclusion of the type I reaction centres (Fig 2.8A). It is also the same the other way around, the type I RC proteins, PsaA, PsaB, PshA and PscA (see Table 2.1) also share a most recent common ancestor excluding the type II RC proteins (Fig 2.8B). Very little sequence identity has been

Fig. 2.8 Phylogenetic trees of type II and type I RC proteins.

(a) Type II RC proteins. A deep dichotomy separates the L and M proteins used in anoxygenic photosynthesis from D1 and D2 used in oxygenic photosynthesis. K represents the ancestral duplication leading to L and M, while D0 represents the duplication leading to D1 and D2. The point denoted **II** represents the most recent common ancestor of all type II RC proteins. **(b)** Type II RC proteins. A deep dichotomy separates PSI, made up by PsaA and PsaB and the homodimeric type I RC proteins. The point denoted **I**, represents the most recent common ancestor of all type I RC proteins. The dichotomies inherent in the evolution of RC proteins demonstrate that oxygenic photosynthesis could be as ancient as anoxygenic photosynthesis and that cyanobacteria are unlikely to have inherited their photosystems from anoxygenic photosynthetic bacteria. The trees were constructed using maximum likelihood methods.

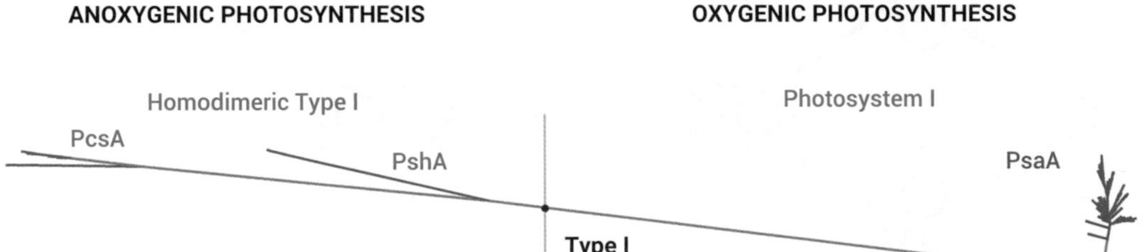

Source: A. W. Larkum, A. W. Rutherford, P. Sánchez-Baracaldo, et. al., 'Early Archean origin of Photosystem II' in *Geobiology*, John Wiley & Sons, 2018, fig. 1a. Licensed under CC BY 4.0. DOI: 10.1111/gbi.12322

detected between type I and type II RC proteins. However, given the highly conserved architectures as revealed by protein structure studies, there is little doubt that they too had a common origin, albeit at an earlier period of their evolution.

During evolutionary studies of photosynthesis the ancestral traits of the earliest RC have been widely debated. For example, did it include F_x and did it have a fused core and antenna? Although proposals have been put forward that the ancestral photosystem was either a type I or a type II, or something in between, no unambiguous evidence yet exists supporting any of these possibilities. What can be inferred unambiguously, however, is that the evolutionary events that led to the functional and structural specialisations of type I and type II RC proteins must predate any other evolutionary transition leading to the different versions of RC proteins inherited by the different groups of bacteria, as illustrated in Fig 2.9. This is true whether they obtained the relevant genes via HGT or vertically. It also means that the ultimate origin of photosynthesis cannot be traced back to any of the known groups of bacteria, as it predates their diversification.

The evolution of type II reaction centres involved two distinct gene duplication events. One leading to D1 and D2 of PSII, and another independent event leading to L and M of the anoxygenic type II RC (Fig 2.8A). Recent analyses of the rates of protein evolution indicate that D1 and D2 of oxygenic photosynthesis evolve a lot slower than L and M of anoxygenic photosynthesis implying that the duplication leading to D1 and D2 occurred long before the one leading to L and M (Fig 2.9). This also means that, although PSII is a heterodimeric RC, it has retained greater symmetry at its core and has a greater number of structural similarities with type I RC but these have been lost in the anoxygenic type II RC. Examples include the retention of antenna domains or the core chlorophyll Z (see Fig 2.3).

Among the anoxygenic groups with type II RCs, the phototrophic species in the Proteobacteria are unique because their photosynthesis genes are present in a single cluster. This gene cluster evolved to allow Proteobacteria to regulate the expression of photosynthetic capacity depending on the presence or absence of oxygen. Hence, under anaerobic conditions the gene cluster is switched on and the organism can sustain photosynthesis. In the presence of oxygen, the cluster is switched off and the bacteria switch to a heterotrophic mode of survival. Phototrophic Proteobacteria inactivate photosynthesis under aerobic conditions because molecular oxygen can react with the excited state of chlorophylls and bacteriochlorophylls generating reactive oxygen species (ROS), such as $^1O_2$ (singlet oxygen), superoxide and peroxide which, in the absence of antioxidants, can cause cell death.

The evolution of a gene cluster in the Proteobacteria has facilitated the lateral transfer of phototrophy to other groups of bacteria. Hence, γ-Proteobacteria and β-Proteobacteria have inherited phototrophy from an ancestral α-proteobacterium. Within α-Proteobacteria, the HGT of photosynthetic gene clusters, as well as losses of these clusters, are well-characterized processes. The same photosynthetic gene cluster has

Fig. 2.9 Schematic representation of the evolution of reaction centre proteins as a function of time.

The section of the tree focusing on PSII and PSI shows the RC gene content of the cyanobacterium, *Chroococcidiopsis thermalis* PCC 7203, as a representative example. These genes are the products of several ancient and more recent gene duplications starting before cyanobacteria (blue dotes at node positions). ChlF, rD1, and CBP represent known paralogues of D1 and CP43 subunits. The tree highlights how the divergence of type I and type II RCs predate the diversification of bacteria, and how it is possible that oxygenic photosynthesis started earlier than traditionally understood. Recent data suggest that the origin of water oxidation started prior to the duplication that led to the origin of CP43/D1 and CP47/D2 and that the earliest type II photosystems were structurally, and potentially also functionally, like water-splitting PSII, but homodimeric. The period of time denoted ΔT represents the span of time between the origin of a homodimeric water-splitting photosystem and the most recent common ancestor of cyanobacteria. On the right side, the evolution of RC proteins is compared with the evolution of cyanobacteria and their closest relatives. Divergence times indicate the relative occurrence of events, rather than exact dates. The vertical transparent bars across nodes of interests highlight the uncertainty of divergence time estimations using molecular clocks.

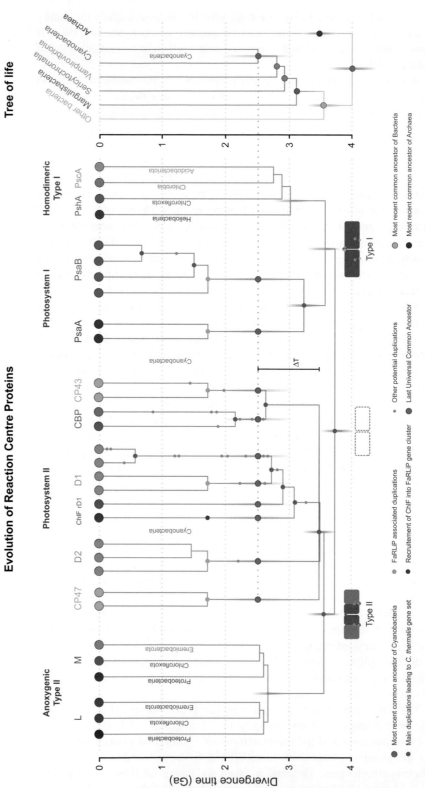

Source: T. Oliver, P. Sánchez-Baracaldo, A. W. Larkum, A. W. Rutherford, T. Cardona, 'Time-resolved comparative molecular evolution of oxygenic photosynthesis' in *Biochimica et Biophysica Acta (BBA) - Bioenergetics*, Elsevier, 2021 1862 (6), fig. 9. Licensed under CC BY 4.0. DOI: 10.1016/j.bbabio.2021.148400

also been transferred via HGT from the Proteobacteria into other phyla. Examples include the Gemmatimonadetes, and possibly into a few strains in other phyla, although the latter are only known from environmental studies and remain poorly characterized in terms of functionality. Recent molecular clock analysis has also suggested that the Chloroflexi obtained photosynthesis via HGT as recently as 1.0 Ga, although no photosynthetic gene clusters that could have facilitated its acquisition are known in this group.

Evidence for independent duplications of the L and M, and of D1 and D2 subunits, has ruled out previous suggestions that cyanobacteria obtained type II RC genes from either Proteobacteria or Chloroflexi. As discussed below, it is thought that water oxidation to oxygen evolved before the duplication that led to D1 and D2. In consequence, given that this duplication occurred before the one leading to L and M, it appears likely that anoxygenic type II RC originated after the origin of oxygenic photosynthesis. This is corroborated by the fact that all anoxygenic type II RCs inherited the LH1 antenna system that is rich in carotenoids (Fig 2.3A). Carotenoids are used to enhance light harvesting and also to quench excited states of chlorophyll that can react with oxygen to generate ROS (see above). Recent genetic and biophysical studies have demonstrated that the configuration of carotenoids found in the LH1 has evolved to sacrifice higher efficiency of excitation transfer in order to optimize photoprotective capabilities. Thus, while anoxygenic type II RCs may appear to be 'primitive', they are actually highly specialized and sophisticated enzymes.

Evolution of type I RCs

The evolution of type I RCs mimics that of type II RCs as described above. This means that the homodimeric type I RCs used in anoxygenic photosynthesis share a most recent common ancestor—to the exclusion of cyanobacterial PSI (Fig 2.8B). This suggests that, at a given point in time, a gene duplication occurred that allowed PSI to become a heterodimer. The question is why did only cyanobacterial PSI become a heterodimer? The answer seems to point towards oxygen.

There are four major groups of anoxygenic phototrophs with homodimeric type I RCs. These are the Heliobacteria, the phototrophic Chlorobi, the phototrophic Acidobacteria, but also the Chloroflexi. The latter is the most recent addition and a significant discovery, because it represents the second group of bacteria to display both type II and type I reaction centres. In 2020 researchers isolated a photosynthetic strain of Chloroflexi with a distinct PscA subunit (Fig 2.8B) capable of assembling homodimeric type I reaction centres, but the genome did not encode a type II RC like the classical and better understood strains. In fact, it was suggested previously that the Chloroflexi had replaced their RC from one type to the other, as the classical strains with type II seemed to have traits that resembled the groups containing type I RCs. However, both type II and I RC in Chloroflexi seem to have ancient histories and are highly divergent relative to their counterparts in other groups, suggesting a complex history of gene transfers, gene replacements, and losses.

The cyanobacterial PSI is the only known heterodimeric type I RC with a core made up by two RC protein subunits known as PsaA and PsaB. It has been suggested that cyanobacterial PSI probably evolved to become a heterodimer as an adaptation to oxygenic photosynthesis. In particular, it is thought that the differences between the subunits allow the organism to fine-tune electron transfer reactions asymmetrically, enabling the diversion of unwanted electron transfer reactions from pathways that could lead to ROS formation during stressful conditions such as high light. In fact, there is evidence indicating that oxygenic photosynthesis had already started when PSI was still a homodimer, and before the duplication leading to PsaA and PsaB. This makes sense because the predicted homodimer would have already acquired protective mechanisms against ROS formation, such as the presence of carotenoids in both PsaA and PsaB to quench excess excitation energy.

In summary, the evolution of RC proteins exhibits an ancient evolutionary bifurcation, which led to the initial emergence of type I and type II RC. This initial divergence involved major structural changes, such as the acquisition of distinct electron acceptors, shifts in the orientation and position of some of the pigments, and rearrangement of the transmembrane α-helices. After that, each RC type bifurcated in parallel into one lineage leading to the photosystems used in anoxygenic photosynthesis, and a second lineage leading to those photosystems used in oxygenic photosynthesis (Figs 2.8 & 2.9). But where do RC proteins originally come from? This is a question that has yet to be answered, not least because RC proteins are unique and have no known relatives, even distant ones, in any other organisms.

Therefore, at present, we can only speculate about the primordial roots of the ancestral RC. One possibility is that ancestral photosystems arose from smaller single-transmembrane polypeptides that trapped photoactive pigments and eventually coalesced into larger protein complexes. Another possibility is that the type II RC originated from ancestral membrane-bound cytochrome b proteins. Yet another proposal is that, because RCs are dimeric, they may have been initially monomeric, and at a later stage evolved into a dimeric state, although it is not clear whether a monomeric photosystem could be photoactive.

2.5   Origin of Photosystem II: the oxygen-evolving system

To fully elucidate how oxygenic photosynthesis originated we must first understand how PSII evolved its unique water splitting chemistry. Central to this is to know how PSII evolved its Mn_4CaO_5 cluster and its highly oxidizing voltage. To answer these questions, we can use the heterodimeric structure of the photosystems to predict the nature of the ancestral states before subunit duplication. That is because, if a trait is conserved in each distinct monomer, it means that it was likely present at the homodimeric stage. Extending this simple rationale, it is possible to reconstruct some, but not all, ancestral traits of the photosystems along their entire evolutionary history as we will now explore.

The midpoint potential for the oxidation of water to O_2 at pH 7.0 is 0.82 V. To achieve water oxidation at ambient temperatures (about 20°C), PSII generates an impressive 1.2 V of oxidizing power. To put this into context, the abiotic thermal splitting of water requires temperatures around 3,000°C. PSII also generates enough power to oxidize the tyrosine, Y_Z, which requires nearly 1.0 V to form the neutral tyrosine radical needed to drive the catalytic cycle of the Mn_4CaO_5 cluster, as described in more detail in Chapter 3. The function and redox properties of Y_Z relies on a hydrogen bond to a nearby histidine residue, H190. The Y_Z residue and its partner H190 in D1 are strictly conserved in all known D1 subunits, and also in all D2 subunits as Y_D and H189 (Fig 2.10). Thus, PSII can generate oxidized tyrosine radicals on both sides of the RC in a symmetrical fashion, even though there is no Mn_4CaO_5 cluster in the D2/CP47 side. The function of Y_D is still not completely understood, but it is thought that it protects the system during assembly of the cluster, plus contributing to its stability in the dark and to the energetic properties of the photochemical chlorophylls. Given that the oxidation of tyrosine requires nearly 1.0 V of oxidizing power, it can be deduced that the homodimeric photosystem already had

Fig. 2.10 Comparison of the electron donor site of D1 and D2.

(a) The Mn_4CaO_5 cluster and associated redox Y_Z-H190 pair plus the inner ligands D170 and E189. **(b)** The electron donor site of D2 showing the redox active Y_D-H189 pair and the phenylalanine residues that occupy positions homologous to D1-D170 and E189. The equivalent position of the Mn_4CaO_5 cluster is shown transparently from an overlap with D1. **(c)** The patch of phenylalanine residues (Fxxx) blocking access to Y_D-H189.

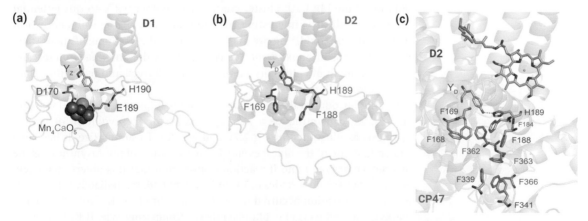

Source: T. Oliver, P. Sánchez-Baracaldo, A. W. Larkum, A. W. Rutherford, T. Cardona, 'Time-resolved comparative molecular evolution of oxygenic photosynthesis' in Biochimica et Biophysica Acta (BBA) - Bioenergetics, Elsevier, 2021 1862 (6), fig. S9. Licensed under CC BY 4.0. DOI: 10.1016/j.bbabio.2021.148400

enough power to split water before the duplication events that led to CP43/D1 and CP47/D2.

Prior to the availability of high-resolution crystal structures, the conserved traits between D1 and D2 were interpreted as meaning that the homodimeric photosystem ancestral to PSII was capable of oxidizing water on both sides of the RC. More recent data from atomic structures show that the CP47/D2 monomeric unit has lost the ability to assemble a catalytic cluster. This is because at every position where coordinating ligands are found in D1 and CP43, hydrophobic and space-filling phenylalanine residues are found in D2 and CP47. These block the access of Mn and water to the cluster (Fig 2.10). There is also a structural rearrangement on CP47 that has modified some of the binding sites of the ancestral D2 subunit. Further evidence that water oxidation originated in a homodimeric system has emerged recently from the structures of homodimeric type I RCs from Heliobacteria and Chlorobi. These homodimeric RCs contain a previously unknown Ca-binding site located in a position equivalent to that of the oxygen evolving complex of PSII with striking structural parallels to the Mn_4CaO_5 cluster (Fig 2.11). Such similarities can only be explained if the earliest photosystems prior to the divergence of type I and type II had already acquired some of the structural elements needed to evolve a catalytic cluster.

How can PSII generate the exceptionally high figure of about 1.2 V of oxidizing power? The most important mechanism is the use of chlorophyll *a* instead of bacteriochlorophyll as a photochemical pigment, due to its higher oxidizing ability. Thus, the midpoint oxidizing potential of chlorophyll *a* in an organic solvent is 0.8 V compared to 0.6 V for bacteriochlorophyll *a*. Other contributing factors are the inherent structural differences between type I and type II RCs, which determine the position of the primary electron donor relative to the transmembrane α-helices and the retention of the antenna domains. In addition, it is thought that the binding of the Mn_4CaO_5 cluster contributes another 0.2 V to this potential in today's PSII. It is interesting to note that many of the traits known to make PSII more oxidizing are ancestral to type II RCs. Some of these traits have been lost in anoxygenic systems, such as the antenna proteins and the replacement of chlorophyll *a* for the more derived and red-shifted bacteriochlorophyll *a*.

Having introduced the evolutionary relationships of type I and II reaction centres, we can introduce a new hypothesis for the evolution of photosynthesis as illustrated in Fig 2.7C and Fig 2.9. To summarize, it has been shown that one of the oldest events in the evolution of photosynthesis is the divergence of type I and II reaction centres, and that this divergence likely occurred before most clades of bacteria diversified. Immediately afterwards, a second bifurcation occurred leading to the split of the RC lineages used in anoxygenic and oxygenic photosynthesis. Comparing type II RCs, it seems that oxygenic PSII is the most likely RC to have retained ancestral traits. In contrast, the anoxygenic type II RC appears atypical, streamlined and to have evolved later. This is evident at a structural level, as seen in Fig 2.3, but it is also consistent with the observation that PSII is the slowest evolving photosystem, and likely one of the slowest evolving enzymes known. In

Fig. 2.11 The Ca-binding site of the homodimeric type I RC.

(a) Full view of PSII showing core proteins (grey) and antenna proteins (orange). **(b)** Full view of the he-liobacterial type I reaction centre showing the core domain (grey) and antenna domain (orange) of PshA. The green spheres are the Ca atoms located symmetrically on each side of the reaction centre. **(c)** Overlap of D1 (orange) and PshA (grey). In PshA, Ca is coordinated by D468, which occupies a structural position similar to Y_Z in D1. Only the 9th and 10th transmembrane helices are displayed for clarity (3rd and 4th in D1). **(d)** Overlap of PsaB of cyanobacterial photosystem I (PSI) (orange), and PshA (grey). The arrows mark the C-terminus of the core proteins. There is no Ca-binding site in PSI as the 11th transmembrane helix of the core domain is about two turns longer (orange arrow). **(e)** Electron density map around the Ca-binding site in Heliobacteria. V608, the C terminal carboxyl group, was modelled in the published structure as a carbonyl, thereby lacking the oxygen that coordinates the Ca atom. **(f)** Close-up of the Ca-binding site show-ing the connection to the antenna domain via N263 and the C-terminus. **(g)** Close-up of the Mn_4CaO_5 cluster highlighting the connection to the antenna via E354. A344, the C terminus, provides a direct ligand to Ca in PSII. **(h)** Scheme of the Ca-binding site of PshA showing the closest distances (A) to residues in the imme-diate vicinity. W stands for water and words in italics highlight structural similarities to PSII.

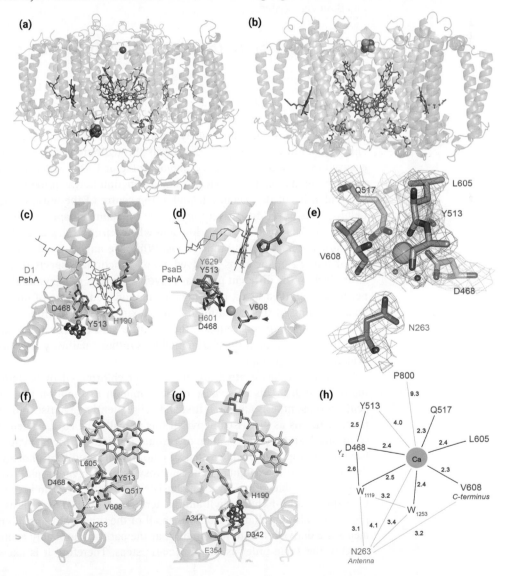

other words, PSII is the most likely RC to have changed the least since the origin of type I and II RCs.

These observations led us and others to challenge the two traditional scenarios shown in Fig 2.3A & B, and to suggest in scenario C that the origin of type I and II RC, at the dawn of photosynthesis, more likely occurred in the context of water splitting to oxygen. If scenario C is indeed correct, it would follow that anoxygenic photosynthesis is a later-evolving by-product of oxygenic photosynthesis, and that the origins of both oxygenic and anoxygenic photosynthesis occurred deep within bacteria, before the evolution of known extant groups. It also implies that many, if not most, extant bacteria, including heterotrophs, are descended from either oxygenic or anoxygenic photosynthetic ancestors. Finally, it also implies that anoxygenic photosynthesis using type I and II RCs in series, as shown in scenarios A and B, never existed.

2.6   Evolution of the chlorophyll biosynthesis pathway

More light can be shed on the evolution of photosynthesis by considering the origins of the chlorophyll biosynthesis pathway. This pathway is related to those of other widely occurring tetrapyrroles, such as heme and cobalamin (vitamin B12) that occur throughout the prokaryotes and eukaryotes, as well as the less common cofactor F430, which is uniquely found in methanogenic archaea. A key distinction between these cofactors is the identity of the metallic ligand at the centre of the tetrapyrrole ring. Chlorophylls have Mg^{2+}, hemes have Fe^{2+}, cobalamins have Co^+ and coenzyme F430 has Ni^+ (Fig 2.12). How and when tetrapyrroles originated and diversified remains poorly understood. While we have a good understanding of the enzymatic steps involved, many functional aspects remain unclear. Each of the different pathways involves dozens of enzymes that have never been studied using phylogenetic methods, and some of our current understanding of this evolutionary complexity has arisen from serendipitous discoveries. In consequence, there is little hard evidence regarding the origin and evolution of the chlorophyll biosynthesis pathway although it is evidently highly complex.

Much debate has centred around whether chlorophyll or bacteriochlorophyll is older. This is because of the common assumption that anoxygenic photosynthesis appeared first and subsequently gave rise to oxygenic photosynthesis as discussed above. If this is really the case, as depicted in scenarios A and B (Fig 2.7), then one would predict that the biosynthesis of bacteriochlorophylls predated that of chlorophylls, and that photosystems that use near-infrared light predated photosystems of oxygenic photosynthesis that use visible light. However, consistent with the scenario in Fig 2.7 C, the central hub of chlorophyll and bacteriochlorophyll biosynthesis is the precursor chlorophyllide *a*, from which all of the other chlorophyllous pigments are derived. It is now thought that the pathway probably evolved in roughly the same order as the enzymatic steps. Therefore it is likely that

Fig. 2.12 Structures of the major modified tetrapyrroles and their relationship to the first shared precursor, uroporphyrinogen III.

The major modified tetrapyrroles shown surrounding the central uroporphyrinogen III include chlorophyll *a*, coenzyme F430, siroheme, and cobalamin. The asymmetrically arranged pyrrole rings in uroporphyrinogen III are named A–D, with the D ring inverted with respect to the other rings. The numbering scheme for the macrocycle is shown for uroporphyrinogen, where positions 1, 2, 5, 7, 10, 12, 15, and 20 are highlighted. In chlorophylls there is a fifth ring, termed ring E, and similarly in F430, there are two extra rings termed E and F. For cobalamin (vitamin B_{12}), the side chains are designated *a–f*. The *X* above the cobalt atom is a cyanide group in vitamin B12; this position is occupied by either a methyl or adenosyl group in the major biological forms of cobalamin.

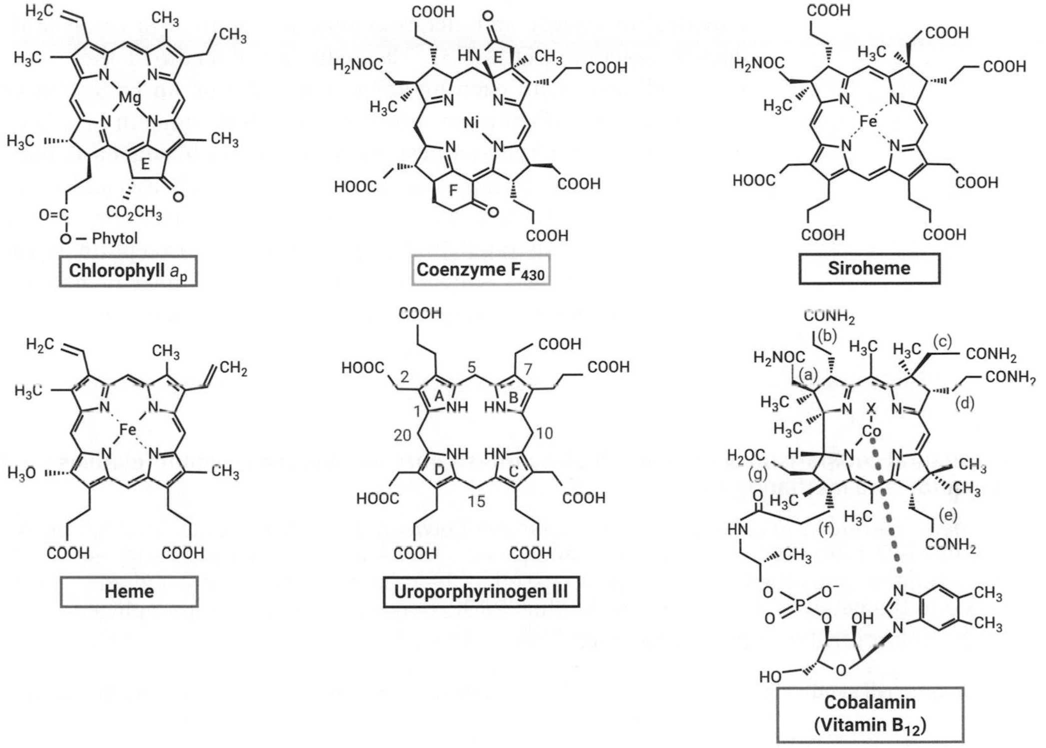

Chlorophyll a_p

Coenzyme F_{430}

Siroheme

Heme

Uroporphyrinogen III

Cobalamin (Vitamin B_{12})

Source: D. A. Bryant, C. N. Hunter, M. J. Warren, 'Biosynthesis of the modified tetrapyrroles - the pigments of life' in Journal of Biological Chemistry, Elsevier, 2020, 295 (20), fig. 1. Licensed under CC BY 4.0. DOI: 10.1074/jbc.REV120.006194

bacteriochlorophyll *a* is the most recent of the major pigments, and that a chlorophyll or chlorophyllide pigment was the ancestral precursor to all the pigments now used by the various extant phototrophs.

All anoxygenic phototrophs with homodimeric type I RCs use a chlorophyll a-derived pigment as the primary electron acceptor. Also, anoxygenic phototrophs that contain chlorosomes, a specialized light-harvesting megacomplex that can synthesize bacteriochlorophylls *c*, *d*, *e*, or *f*, which despite their name, are indeed chlorophylls. Therefore, it can be concluded that the capacity to synthesize a chlorophyllide precursor had already evolved before the events leading to the diversification of the major groups of phototrophic bacteria. Two key enzymes in this pathway are not only homologous to nitrogenase, which is used by microorganisms to fix nitrogen gas (N_2) directly from the atmosphere into ammonia, but also to Ni^{2+}-sirohydrochlorin a,c-diamide reductase. The latter is an enzyme required to make the coenzyme F430 that is exclusively found in methanogenic archaea. Coenzyme F430 is a cofactor of methyl-coenzyme M reductase, which is the key enzyme responsible for biological methane production, catalysing the last methane-release step in the process.

This family of enzymes is structurally characterized by having a reductase domain and a catalytic domain, in which the reductase domain provides the energy and electrons that are required to catalyse the reductive reactions (Fig 2.13). The reductase domain in these enzymes is homodimeric with each monomer coordinating an Fe_4S_4 cluster and containing an ATP-binding motif. The catalytic domain is a heterodimer in the chlorophyll synthesis enzymes and in nitrogenase, but a homodimer in the Ni-tetrapyrrole reductase. The fully assembled enzyme combines two reductase domains and two catalytic domains making an octamer as illustrated in Fig 2.13. The only exception is the Ni-tetrapyrrole reductase which is made up of a tetramer of homodimers instead of octamers. Nitrogenase in turn is homologous to an assembly

Fig. 2.13 Structural comparison of chlorophyll biosynthesis enzymes and their distant relatives used in nitrogen fixation and methanogenesis.

This figure demonstrates the close structural similarities between the enzymes of chlorophyll biosynthesis with those of nitrogen fixation and methanogenesis. These structural affinities suggest possible evolutionary relationships these proteins as discussed in the main text. The reductase domain is shown in red/pink and the catalytic domains in green/blue. DPOR, dark-operative protochlorophyllide oxidoreductase; COR, chlorophyllide a oxidoreductase.

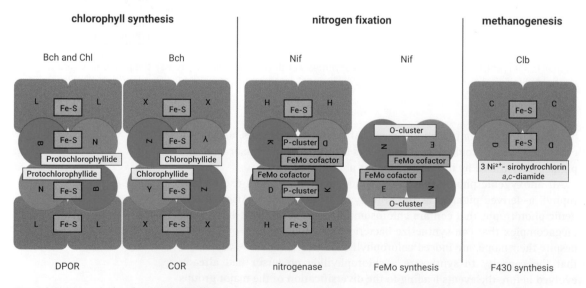

factor called NifEN that is required for the biogenesis of the complex Fe_7MoS_9C catalytic cluster of nitrogenase that is a tetramer assembled from catalytic domain dimers. This enzyme is found in most organisms capable of nitrogen fixation.

While there is a great deal of structural similarity between these enzymes, they are only distantly related. This makes phylogenetic reconstruction challenging and inconclusive. Overall, however, DPOR and COR (see Fig 2.13) share slightly greater sequence similarity, as do the methanogenesis enzyme and nitrogenase, particularly between their reductase domains. While DPOR and COR likely share a more recent common ancestor to the exclusion of all other enzymes, it cannot be concluded that one predated or gave rise to the other. Therefore, it is unclear whether the ancestral enzyme to DPOR and COR acted on protochlorophyllide or another Mg-tetrapyrrole precursor during the early evolution of the chlorophyll biosynthesis pathway. On the other hand, if nitrogen fixation originated in methanogenic archaea, as suggested by some researchers, the evolution of this large family of enzymes would suggest an ancient bacteria/photosynthesis and archaea/methanogenesis split. This is consistent with geochemical evidence indicating that both processes were already occurring around 3.4 Ga.

2.7   The rise of the cyanobacteria

The origin of oxygenic photosynthesis and of cyanobacteria are generally considered to be the same thing. Therefore, the question 'when did oxygenic photosynthesis originate?' is often taken to mean 'when did cyanobacteria originate?' However, an important and often overlooked point is that all cyanobacteria inherited a 'fully evolved' oxygenic photosynthesis. That is to say that all cyanobacteria share a highly conserved heterodimeric PSII and PSI with a complete set of structural components and light harvesting systems. Therefore, the most recent common ancestor (MRCA) of cyanobacteria was already a very sophisticated oxygenic phototroph that had photosystems virtually indistinguishable from those in extant cyanobacteria (Fig 2.9). Despite this high degree of conservation modern cyanobacteria have evolved numerous ways to fine tune the efficiency of their photosystems as described in Bigger Picture 2.1.

Bigger picture 2.1
Evolution never stops

The Photosystem II complex is highly conserved across all organisms capable of oxygenic photosynthesis. For this reason it is often suggested that PSII is a 'frozen metabolic accident', meaning to say that it could have only evolved that way and that it is more or less immutable. There is a grain of truth in this notion as we see that the PSII of today's most recently evolved land plants is still very similar in its overall organization to that of cyanobacteria, from which the ancestors of plants diverged over two billion years ago. However, more recent evidence has also revealed that cyanobacteria have evolved an impressive array of mechanisms to optimize their responses to changing light conditions. Strikingly, their genomes contain additional but modified copies of photosystem subunits that can be swapped around as needed, enabling them to fine-tune the properties of the photosystems in response to new conditions as discussed in Chapter 7.

The best-known adaptation is that which involves a change in D1 subunit, with most strains encoding five or six versions of D1. Thus, under high-light conditions, a strain might swap one D1 for a slightly different D1 that is better optimized to work under high-light intensities. Other specialized D1 subunits are thought to have evolved to 'switch-off' PSII during conditions when oxygenic photosynthesis is not needed. Many other forms of D1 still remain to be studied. Similar strategies are also used to optimize PSI by a wide range of cyanobacteria with known species having the potential to assemble up to eight different versions of PSI. We are only just beginning to become aware of this huge molecular diversity and still do not fully understand how, when and why the different variants are expressed (Fig 2.9).

Today, scientists are looking at this remarkable and hitherto unknown natural diversity of cyanobacteria for inspiration for efforts to re-engineer photosynthesis using biotechnological strategies. For example, it is thought that crops could be enhanced by engineering photosystems capable of better using the far-red end of the visible light spectrum, which is not used by most algae and plants. This is an important part of addressing the challenges of climate change and food security where photosynthesis plays many vital roles. Examples of the benefits of such research include the generation of high-value products, the enhancement of crop yields, or even the engineering of novel devices to perform hybrid biological/chemical or even totally artificial forms of photosynthesis as discussed in Chapter 7.

Water-splitting chemistry is thought to have started with homodimeric PSII and PSI RCs. In other words, there was a period of time between the origin of oxygenic photosynthesis, when water oxidation started for the very first time in homodimeric systems, and that of the MRCA of cyanobacteria that inherited fully capable and optimized heterodimeric photosystems. This period of time, denoted ΔT in Fig 2.9, was the interval between the origin of oxygenic photosynthesis and the diversification of cyanobacteria from their MRCA. But it is unclear whether ΔT was a few million years or hundreds of millions of years. This issue can now be addressed using genomic and metagenomic approaches that have already greatly improved the understanding bacterial diversity.

Cyanobacteria are now known to be closely related to a newly discovered and as yet uncultured bacterial, the Vampirovibrionia. These were originally found in metagenomes of the gut in humans and other animals. They are heterotrophic and can be either symbionts with animals or predators of photosynthetic eukaryotes, while other species are present in aquifers where light cannot penetrate. Two other bacterial groups, Sericytochromatia and Margulisbacteria, are also closely related to cyanobacteria although no photosynthetic representatives have yet been found (Fig 2.9). The Sericytochromatia are enigmatic as only a handful of environmental genomes are known, while the Margulisbacteria is a hugely diverse group discovered in both marine and terrestrial environments. Surprisingly, the few characterized Margulisbacteria live in complex symbiotic associations involving several interacting microbial partners within the gut of termites or as specialized endosymbionts of early-branching animals, like the placozoa.

The discovery of non-photosynthetic relatives of cyanobacteria has led to the proposal that oxygenic photosynthesis could have evolved soon after the divergence of Vampirovibrionia by the acquisition of photosynthesis genes via HGT. Comparative analyses of the predicted gene content of Margulisbacteria suggested that their MRCA was non-photosynthetic and dependent on hydrogen or fermentative metabolism. Molecular clock analyses, including sequences from Vampirovibrionia and cyanobacteria, further suggest that the time span between their divergence could have been several hundred million years, implying that oxygenic photosynthesis had to have emerged relatively rapidly. However, molecular clocks applied to PSII indicated that water oxidation predate the MRCA of cyanobacteria by well over a billion years. This is consistent with an RC evolution scheme where the photosystems origins predate the diversification of most groups of bacteria. Indeed, recent comparative evolution studies of PSII indicate that it has molecular evolution patterns usually attributed to the oldest enzymatic systems such as ATP synthase, RNA polymerase, or the ribosome.

These findings imply that the Margulisbacteria, Sericytochromatia and Vampirovibrionia, and perhaps other bacteria, each diversified following the loss of primordial forms of oxygenic photosynthesis on multiple occasions over geological time to become specialist heterotrophic symbionts of other bacteria and eukaryotic organisms. While the best accepted cyanobacterial fossils date to 2.0 Ga, cyanobacteria-like microbial mats and stromatolites exist from before 3.8 Ga (see Chapters 1 and 5). Although there is an emerging consensus that oxygenic photosynthesis existed by 3.0 Ga, it is unclear whether this was due to the activity of cyanobacteria with known living descendants or from ancestral forms of bacteria that predated the diversification of extant clades and even cyanobacteria themselves.

If oxygenic photosynthesis originated more than a billion years before GOE1 why did the Earth not become oxygenated before this date?

As highlighted in Chapter 1, the answer to this question remains unclear. For gaseous oxygen to accumulate, its rate of production via photosynthesis must be greater than its consumption by respiration due breakdown of dead organic matter or its absorption by highly reducing minerals such as iron deposits. If both processes are balanced, no net oxygen production occurs. But rates of oxygenic photosynthesis in turn depend on the availability of nutrients required to sustain large communities of photosynthetic bacteria. Nutrient availability depends on geological factors such as the size and shape of the continents. Whether oxygen accumulates in the ocean and the atmosphere also depends on their composition. For example, a highly reducing atmosphere would rapidly scavenge and remove any released oxygen. Other factors such as the rate at which hydrogen escaped into space might have contributed to the rates of oxygen accumulation. Recent studies suggest that even if oxygenation of the Earth started around 4.0 Ga, no particular triggers or evolutionary innovations are needed to explain the step-wise oxygenation of the planet, and this could be the result of the inherent long term nutrient biogeochemical cycles.

Chapter summary

- Resolving the origin and early evolution of photosynthesis is one of the greatest challenges of evolutionary biology, second only to the origin of life itself. The past decade has seen great advances in our understanding of how, where and when the various forms of photosynthesis started.
- Photosynthesis is an ancient process that probably originated in bacteria during some of the earliest events following the emergence of cellular life.
- Bacterial photosynthesis can be either oxygenic or anoxygenic but it is not yet clear which form came first. However, emerging evidence suggests that oxygenic photosynthesis may have arisen first, possibly in the early Archean Eon, and that anoxygenic photosynthesis is a derived process that evolved soon afterwards.
- Bacterial photosynthesis is a highly dynamic process with many different forms. These bacteria have continued to innovate their photosynthetic mechanisms to adapt to dramatic climatic and environments for at least 2 to 3 billion years.
- By studying these processes and adaptations, new ideas should emerge to help tackle some of the formidable sustainability and climatic challenges that we currently face as a global community.

Further reading

Bryant DA, Hunter CN, Warren MJ (2020) Biosynthesis of the modified tetrapyrroles—the pigments of life. *J Biol Chem* 295, 6888–6925. DOI: https://doi.org/10.1074/jbc.REV120.006194

Comprehensive review of tetrapyrrole biosynthesis including chlorophylls and bacteriochlorophylls.

Cardona T et al (2019) Early Archean origin of Photosystem II. *Geobiology* 17, 127–150. DOI: https://www.ncbi.nlm.nih.gov/pmc/articles/PMC6492235/
An in-depth analysis of the evolution of PSII as a function of time.

Chen JH et al (2020) Architecture of the photosynthetic complex from a green sulfur bacterium. *Science* 370, eabb6350. DOI: 10.1126/science.abb6350
The first atomic resolution structure of the green sulphur bacterial reaction centre.

Gisriel C et al (2017) Structure of a symmetric photosynthetic reaction center-photosystem. *Science* 357, 1021–1025. DOI: https://science.sciencemag.org/content/357/6355/1021.long
The first crystal structure of a homodimeric type I RC from Heliobacteria.

Hohmann-Marriott MF, Blankenship RE (2011) Evolution of photosynthesis. *Ann Rev Plant Biol* 62, 515–548. DOI: 10.1146/annurev-arplant-042110-103811
A well-balanced review showcasing the diversity of photosynthetic bacteria.

Jordan P et al (2001) Three-dimensional structure of cyanobacterial Photosystem I at 2.5 Å resolution. *Nature* 411, 909–917. DOI: 10.1038/35082000
The first structure of cyanobacterial Photosystem I.

Oliver T et al (2020) Time-resolved comparative molecular evolution of oxygenic photosynthesis. *BBA Bioenergetics* 1862, 1–20. DOI: https://doi.org/10.1101/2020.02.28.969766
Phylogenetic and structural analyses suggesting that oxygenic photosynthesis originated close to the origin of life.

Orf GS, Gisriel C, Redding KE (2018) Evolution of photosynthetic reaction centers: insights from the structure of the heliobacterial reaction center. *Photosynth Res* 138, 11–37. DOI: https://doi.org/10.1007/s11120-018-0503-2
Using the structure of the homodimeric type I RC to shed light on the evolution of photosynthesis.

Thiel V, Tank M, Bryant DA (2018) Diversity of chlorophototrophic bacteria revealed in the omics era. *Ann Rev Plant Biol* 69, 21–49. DOI: https://doi.org/10.1146/annurev-arplant-042817-040500
A well-balanced review of our increased understanding of the diversity of photosynthetic bacteria as facilitated by environmental studies.

Umena Y et al (2011) Crystal structure of oxygen-evolving Photosystem II at a resolution of 1.9 Å. *Nature* 473, 55–60. DOI: https://doi.org/10.1038/nature09913
The first structure of Photosystem II that allowed positional resolution of all of the atoms of the Mn_4CaO_5 cluster.

Yu LJ et al (2018) Structure of photosynthetic LH1-RC supercomplex at 1.9 Å resolution. *Nature* 556, 209–213. http://ousar.lib.okayama-u.ac.jp/en/56313
A high resolution atomic structure of the RC from a purple bacterium.

Discussion questions

2.1 Is there a type of photosynthesis that can be described as primitive? Why?

2.2 What are the three main different scenarios that describe the evolution of photosynthesis and what are their strengths and weaknesses?

2.3 What would be the implications for the history of life on Earth that oxygenic photosynthesis started close to the beginnings of cellular and bacterial life?

3 EUKARYOTIC PHOTOSYNTHESIS

Learning objectives

- Exploring the molecular architecture of photosynthetic membranes.
- Understanding the mechanisms of light to chemical energy conversion in oxygenic photosynthesis.
- Describing examples of different light-harvesting complexes in cyanobacteria and photosynthetic eukaryotes.
- Providing a molecular overview of water oxidation and oxygen evolution in PSII.
- Summarizing the function of cytochrome $b_6 f$ complex.
- Discussing the functional differences between PSI and PSII.
- Understanding how photosynthetically produced ATP and NADPH are used for **CO_2 fixation**.
- Critically analysing the need for carbon concentration mechanisms for efficient photosynthesis.

3.1 Introduction

Oxygenic photosynthesis is by far the most important form of photosynthesis on Earth. It accounts for an estimated 3000-fold greater amount of carbon fixation compared to all of the various forms of anoxygenic photosynthesis. As we have seen in Chapter 2, it is unclear what the earliest forms of photosynthesis were like. There is emerging evidence that oxygenic photosynthesis might have its origins close to the beginnings of cellular life, around or before 4 Ga, but this remains to be confirmed. However, oxygenic photosynthesis was the only form that was acquired by eukaryotes following a unique endosymbiotic event between a eukaryotic heterotroph and a cyanobacterium (see Chapter 4).

Over the past two billion years the basic chemistry of oxygenic photosynthesis in algae and plants has changed remarkably little.

It remains nearly identical to the original mechanism that had been previously employed by cyanobacteria prior to the endosymbiotic event with a eukaryotic cell that occurred about or before 2.0 Ga. In both cyanobacterial and eukaryotic photosynthesis, photons of light energy, mostly in the region 400 to 700 nm, are intercepted by large arrays of light-harvesting, pigment-protein complexes housed in or on thylakoid membranes. These pigment-protein complexes contain numerous tightly organized chlorophyll molecules, plus accessory pigments such as carotenoids and xanthophylls. The light-harvesting complexes then pass on their absorbed energy, via a quantum process of resonance transfer, to the two reaction centre complexes, namely PSI and PSII, which are the actual sites of the photochemical reactions.

The channelling of light-derived energy (excitation energy) to the reaction centre of PSII enables molecules of water to be split to create protons and electrons plus oxygen gas. In algae and plants, the two reaction centres are linked by a third iron-containing protein complex, called cytochrome b_6f, to form a chain of electron carriers leading all the way from water to NADPH.

Fig. 3.1 Simplified Z-scheme of photosynthesis.

At the bottom left, light absorption by the redox chlorophylls of PSII, also known as P680, triggers the formation of the very reducing P680* excited state, initiating charge separation. The formation of the oxidized P680 can generate over +1.2V of oxidizing power required to split water. The electrons extracted from water are transported to the cytochrome b_6f complex via a pool of plastoquinone molecules in the thylakoid membrane (PQ). The cytochrome b_6f complex subsequently oxidizes the electron carrier protein plastocyanin (PC). Light absorption by the redox chlorophylls of PSI, generally referred to as P700, triggers the formation of the very reducing P700* state, initiating charge separation in this complex and culminating in the reduction of NADP⁺ via a series of PSI bound cofactors and the electron shuttles ferredoxin (Fd) and ferredoxing-NADP⁺ reductase.

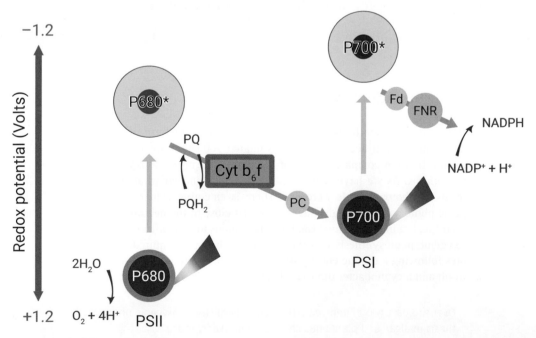

This electron transport pathway, from PSII to cytochrome b_6f and on to PSI, is widely known as the Z-scheme. Its inputs are light energy and water and its products are protons, NADPH, and oxygen. A fourth protein complex, ATP synthase, uses the proton gradient generated by electron transfer through the thylakoid membrane to catalyse ATP formation. All of the light-using reactions occur on membrane-bound protein complexes located within the thylakoid membranes of cyanobacteria or the chloroplast. As discussed in Chapter 7, advances in our knowledge of the mechanism of photosynthesis could eventually lead to the development of new forms of biologically derived solar-powered devices. This chapter is mainly focused on the mechanisms of oxygenic photosynthesis in eukaryotes, ie algae and plants, although there is some comparative analysis of the extant cyanobacterial mechanisms that can inform us about the origins of this fascinating process that has had such a momentous effect on biological evolution.

3.2   Ultrastructural context of oxygenic photosynthesis in thylakoid membranes

Oxygenic photosynthesis in both cyanobacteria and eukaryotes takes place on and in specialized membranes called thylakoids. In cyanobacteria, the thylakoids are pairs of bilayers that enclose a space physically separated from the cytosolic medium. In this enclosed space, called the lumen, the pH is lower than in the outer aqueous phase known as the stroma. The best characterized thylakoids in some model species of cyanobacteria appear as multiple concentric layers around the interior of the cell, but there is a wide and largely uncharacterized diversity of membrane morphologies. In photosynthetic eukaryotes the thylakoids are part of the chloroplast membrane system. As shown in Fig 3.2, a typical higher plant chloroplast contains a series of thylakoid membranes, some of which are stacked to form grana while others link adjacent granal stacks.

> Thylakoid membranes are the sites of light-harvesting and electron transport reactions that generate reductive power in the form of NADPH. The proton gradient established as part of photosynthetic electron transport also powers the formation of ATP.

The NADPH and ATP then enable chloroplasts to carry out other important metabolic functions. These include conversion of simple inorganic precursors, such as CO_2, nitrogen, phosphorus, and sulphate, into complex organic molecules, such as lipids, nucleotides, carbohydrates, pigments, amino acids, and isoprenoids.

The three protein complexes that form the core of the Z-scheme of electron transport, namely PSII, cytochrome b_6f, and PSI are shown in their membrane context in Fig 3.3. During electron transport, electrons from PSII and protons from the stromal side of the thylakoid are passed onto the lipophilic molecule, plastoquinone, which is able to migrate through the hydrophobic core of

Fig. 3.2 Typical higher plant chloroplast and some key metabolic features.

The chloroplast takes simple low-energy components (left box) and converts them into more complex high-energy metabolites (right box) via photosynthesis. The chloroplast is surrounded by a double membrane made up of an inner and outer envelope. The main aqueous space inside the chloroplast is called the stroma and is the site of most of the metabolism apart from the photosynthetic reactions. The thylakoid membranes are made up of flattened vesicles that enclose an inner aqueous space called the lumen. Thylakoids can be stacked on top of one another to form grana, or remain unstacked and connect adjacent granal stacks. All of the light reactions occur in thylakoid membranes with the water-splitting PSII and cytochrome b_6f complexes mainly located in stacked regions while PSI and ATP synthase complexes mainly located in unstacked regions.

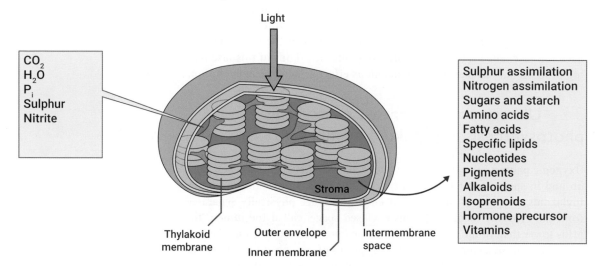

the lipid bilayer until it reaches the cytochrome b_6f complex. The plastoquinone then releases electrons to cytochrome b_6f, while protons are released on the lumen side of the membrane as discussed below. Once the electrons have passed through the cytochrome b_6f complex, they are transferred one-by-one to the soluble carrier, plastocyanin. This carrier travels through the aqueous lumen until it reaches PSI and donates electrons to it. In the final stage of electron transport, PSI uses light energy to provide additional power that enables it to pass on electrons to $NADP^+$, thus forming NADPH.

The release of protons on the lumen side of the membrane, as described above, establishes a steep proton gradient across the thylakoid membrane. This results in the stromal space having an alkaline pH of about 8.0 while the lumen has an acidic pH of about 4.5. The resulting pH difference (ΔpH) of 3.5 units creates a light-induced transmembrane proton gradient across the thylakoid membrane that can generate a proton-motive force of 0.20 V or a ΔG of -4.8 kcal mol^{-1} (-20.0 kJ mol^{-1}). This proton gradient provides sufficient energy to drive the conversion of ADP to ATP, as the protons flow back down their concentration gradient via a channel inside the ATP synthase complex and back out into the stroma space. The ATP product of this photophosphorylation process, together with the highly reducing compound, NADPH, is then used to power the conversion of CO_2 into carbohydrates and other important reactions as described above.

Fig. 3.3 Light-dependent reactions of photosynthesis at the thylakoid membrane.
(a) The three major protein complexes involved in electron transport are PSII, cytochrome b_6f and PSI, while the ATP synthase complex uses the proton gradient generated during electron transport to generate the formation of the high-energy compound, ATP. **(b)** Thylakoid membranes form sealed flattened sacs that enclose an aqueous space called the lumen. During electron transport, protons are pumped from the external stromal compartment into the lumen. This gives rise to a proton gradient of about 3.5 pH units. When protons flow back down this gradient via proton pathways in the ATP synthase complexes, the resultant energy can be used to power ATP synthesis.

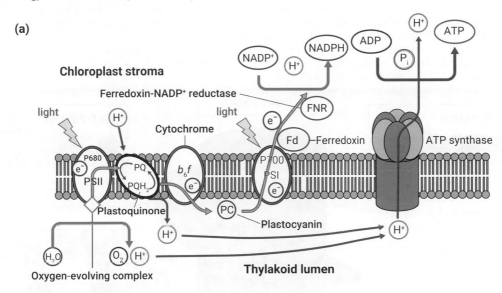

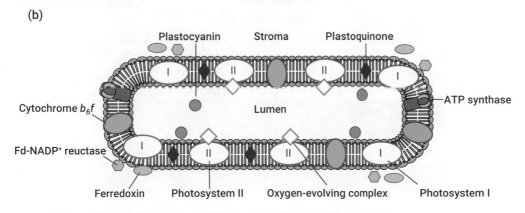

Source: Redrawn from Somepics, 2015 / Wikimedia Commons / Public Domain

The ultrastructure of the thylakoid membranes in a typical higher plant chloroplast is shown schematically in Figs 3.4 and 3.5. While some of the thylakoid membranes are stacked on top of one another to form grana, other thylakoids remain unstacked. Each thylakoid comprises a sac-like structure that is equivalent to a flattened vesicle. The proportion of stacked and unstacked thylakoid membranes varies according to

Fig. 3.4 Helical model of thylakoid membrane organization and putative location of photosynthetic protein complexes involved in linear electron transfer.

Higher plant thylakoid membranes are folded into layers of appressed grana stacks (green). The edges of the grana thylakoids are termed the curvature area. The space between adjacent thylakoid layers in grana is the stromal gap, in which negatively and positively charged amino acid residues (blue and red dots, respectively) of neighbouring LHCII trimers interact and where cations, such as Mg^{2+} (orange dots), are screened between the negatively charged stromal loops of adjacent LHCII trimers. The separate grana stacks are helically connected by four to six unstacked thylakoids (blue). The interface between the appressed and non-appressed membranes is called the granal margin (purple). Linear electron flow (blue arrows) between the photosystems produces NADPH, while protons are pumped from the stroma into the thylakoid lumen (dashed red arrows). The consequent proton motive force powers the synthesis of ATP. The protein complexes are unevenly distributed in the thylakoid membrane: the PSII–LHCII and cytochrome b/f complexes are mainly found in the grana, while the large PSI–LHCI and ATP synthase reside in the non-appressed thylakoids and grana margins.

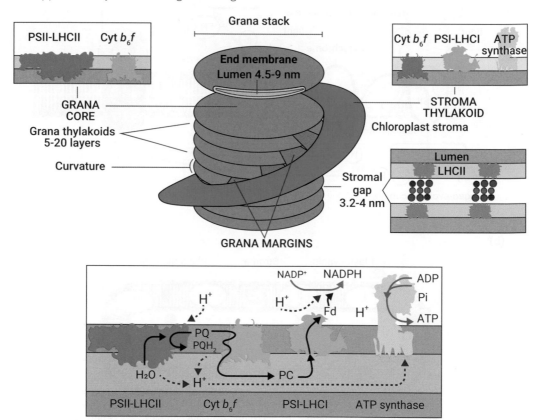

M. Rantala, S. Rantala, E. Aro, 'Composition, phosphorylation and dynamic organization of photosynthetic protein complexes in plant thylakoid membrane' in Photochem. Photobiol. Sci, RSC, 2020, 19, 604-619, fig. 1. Licensed under CC BY 3.0. DOI: 10.1039/D0PP00025F

cell type and environmental conditions. The stacked membranes are the major sites of PSII and its associated LHCII complexes. Each PSII complex is surrounded by a variable number of LHCII complexes to form a 'super-complex'. These PSII–LHCII super-complexes are located in the stacked granal regions and electrostatic interactions between protruding LHCII complexes on adjacent thylakoids are responsible for keeping the stacks intact.

Fig. 3.5 Arrangement of the major photosynthetic protein complexes in stacked (appressed) and unstacked (non-appressed) thylakoid membranes.

This diagram shows the outline structures of the various photosynthetic protein complexes at the same relative scale. The transmembrane helical regions that anchor the proteins into the thylakoid membranes are especially prominent. Stacked or appressed thylakoids are only 2nm apart while the lumen space is about 21 nm from top to bottom. The lower part of the diagram summarises the direction of operation of photosynthetic electron transport which proceeds from PSII to cytochrome b_6f in stacked thylakoids to PSI in unstacked thylakoids where the ATP synthase complex is also located.

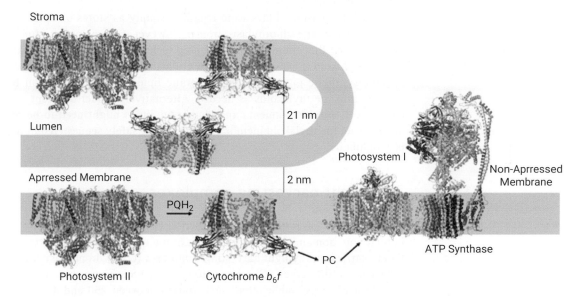

Source: S. Bhaduri, S. K. Singh, W. Cohn, S. S. Hasan, J. P. Whitelegge, W. A. Cramer, 'A novel chloroplast super-complex consisting of the ATP synthase and photosystem I reaction center' in PLoS One, 15(8), 2020, fig. 1. Licensed under CC BY 4.0. DOI: 10.1039/D0PP00025F

Under certain environmental conditions, the protruding surface regions of LHCII become phosphorylated, mainly at conserved serine and threonine residues. This leads to repulsion between adjacent thylakoid membranes and a reduction in the overall amount of granal stacking. Cytochrome b_6f complexes are also mainly found in the stacked granal regions of thylakoids, indicating that the processes of water-splitting and subsequent electron transfer all the way to plastocyanin mostly occur in the grana. The unstacked, or non-appressed, thylakoid regions are enriched in the remaining components of photosynthetic electron transport, namely PSI and its adjacent NADPH-producing carriers. There is also recent evidence that the thylakoid ATP synthase is associated with PSI and LHCI to form another 'super-complex', as shown in Fig 3.5. The reason for the association of the ATP synthase with PSI has yet to be determined but it is possible that it facilitates the localized coupling of the proton gradient generated in PSI with ATP synthesis.

Thylakoids are highly dynamic membranes that need to adjust their photosynthetic activities on timescales ranging from less than a second to many weeks. This is necessary in order for them to adapt to different light regimes and other environmental factors, including temperature and day length. As discussed above, rapid changes in granal stacking can enable thylakoids to adapt to sudden short-term changes in illumination, for example between intermittent periods of full sunlight and various degrees of transient cloud cover. Longer-term changes can be achieved by adjusting the protein and lipid content of the thylakoid membranes. One recently discovered mechanism involves the participation of lipidic droplets called plastoglobules. Plastoglobules are spherical droplets containing a core of neutral lipids surrounded by a galactolipid monolayer and a specific population of proteins.

Previously, plastoglobules were regarded simply as stores of lipid that accumulated in the chloroplast stroma. However, several recent studies have revealed that these droplets are able to fuse with the edge, or margin, of thylakoid membranes. The fused plastoglobules contain a variety of photosynthetic pigments and membrane lipid precursors that can be released into the thylakoids as needed. Alternatively, under different circumstances, surplus pigments and membrane lipids might need to be removed from the thylakoid membranes and can be safely sequestered in the plastoglobules.

3.3   Light harvesting complexes

The photochemical processes in the reaction centres of PSI and PSII occur in their core domains (see Chapter 2). Light interception and energy transfer are carried out by light harvesting complexes (LHCs). In photosynthetic eukaryotes LHCI serves PSI while LHCII serves PSII. These two LHCs are huge molecular assemblies that each contain between 250 and 400 chlorophylls plus accessory pigments such as carotenoids. The LHCs intercept light over a broad range of wavelengths and are particularly important in maintaining a steady flow of energy to their respective reaction centres under fluctuating light conditions. For example, even during full sunlight, each chlorophyll molecule can only absorb up to about ten photons of light per second, which is much slower than the capacity of the reaction centres to use light energy. Therefore, in the absence of specialized LHCs, the reaction centres would remain inactive most of the time. Following light capture by an LHC, there is a highly efficient and rapid energy transfer process whereby within a fraction of a nanosecond, the absorbed light energy (excitation energy) is channelled to the reaction centre, where charge separation occurs (see Chapter 2).

A photon is a quantum entity existing as a combination of a particle and a wave that contains a discrete amount of energy as defined by its wavelength.

Within the visible part of the electromagnetic spectrum the wavelength of a photon corresponds to its perceived colour. Hence, lower wavelength (about 400nm) photons are perceived as blue light while higher wavelength (about 700nm) photons are perceived as red light. When a pigment molecule absorbs the energy of a photon, the energy of the photon is transferred to the pigment, which enters an excited state. This excitation energy can be transferred very rapidly and with near 100% efficiency, which does not involve the transfer of electrons or redox reactions. Thus, the pigments in a light-harvesting complex are organized somewhat like a funnel where there is a directionality of energy transfer towards the photosystem's photochemical core. This is achieved by cofactor-protein complexes that contain a high density of tightly packed and optimally positioned pigment cofactors, often in direct physical contact with each other. These LHCs are arranged to be energetically fine-tuned to absorb light at a higher-energy (shorter wavelengths or 'bluer') state than that of the photosystem, which is activated with red photons.

Classically, light harvesting is described as the excitation energy 'hopping' from one pigment of higher-energy to another with a lower-energy state. Extra-large LHCs, such as phycobilisomes (Fig 3.6) are found in cyanobacteria and non-green algae, and collect photons of shorter wavelengths at their periphery and of longer wavelength in regions closer to the reaction centres. In green algae and plants, phycobilisomes are replaced by smaller LHCs with similar functions. With every excitation hop or transfer, a small amount of energy is lost as heat, ensuring that the excitation moves directionally down an energy gradient towards the photochemical core. It is however not strictly necessary for excitation transfer to occur 'downhill', as the thermal energy of the environment can sometimes push the excitation energy of the pigment 'uphill'.

This phenomenon of 'uphill' energy transfer has several uses as part of strategies to protect the sensitive reaction centres from over-excitation or to improve light harvesting in the less energetic redder regions of the spectrum. Because excitation transfer occurs at the level of photons and atoms, it cannot be entirely described by classical physics, and quantum effects occur. This can sometimes make light-harvesting appear as a counterintuitive process. For example, the excitation of photon absorbed by a pigment may end up localized in a different pigment, but the exact pathway the excitation energy took to get there cannot be exactly deciphered, and in some sense the excitation follows several paths simultaneously.

In contrast to the highly conserved photosystems, there are many different types of light-harvesting systems that have multiple evolutionary origins. Indeed, LHC diversity is as extensive as that of the various photosynthetic lineages. In particular, oxygenic organisms can sometimes contain more than one type of complex, and these are dynamically optimized depending on light intensity, light quality, and the resources available to the organism. An example is phycobilisomes, which are massive cytosolic protein complexes, each containing hundreds of protein subunits and thousands of pigment molecules. As shown in Fig 3.6, a phycobilisome is typically arranged as a series of cylindrical structures consisting of a core linked to a number of peripheral rods. The core part of the complex is linked to PSII and is optimized to deliver light to its reaction centre. However, because the entire complex is so large, multiple PSI and PSII complexes can be found underneath a single phycobilisome. The number of core cylinders varies from organism to organism but usually ranges from two to five. The composition of the peripheral rods and core subunits and pigments is dynamic and can be changed and optimized depending on the quality of light.

The most recent common ancestor of photosynthetic eukaryotes, ie the progenitor of all of the Archaeplastida including primary algae and plants

Fig. 3.6 The phycobilisome (PBS).

(a) Side view of the PBS from *Synechococcus* sp. PCC 7002. The cryo-EM map is displayed in surface representation, with the core and peripheral rods are shown in different colours. Allophycocyanin trimer, green; phycocyanin, pink, sky blue, and yellow. **(b)** Bottom view of the PBS. ApcF, ApcD, and L_{CM} are highlighted in orange, magenta, and red, respectively. **(c)** Distribution of linker proteins in the PBS. Models of linker proteins are shown in cartoon representation and separately coloured. Linker protein L_{CM} binds the PBS to PSII. **(d)** Schematic model of the PBS protein subunits. **(e)** to **(h)** Same as **(a)** to **(d)**, but for the PBS from *Nostoc* sp. PCC 7120.

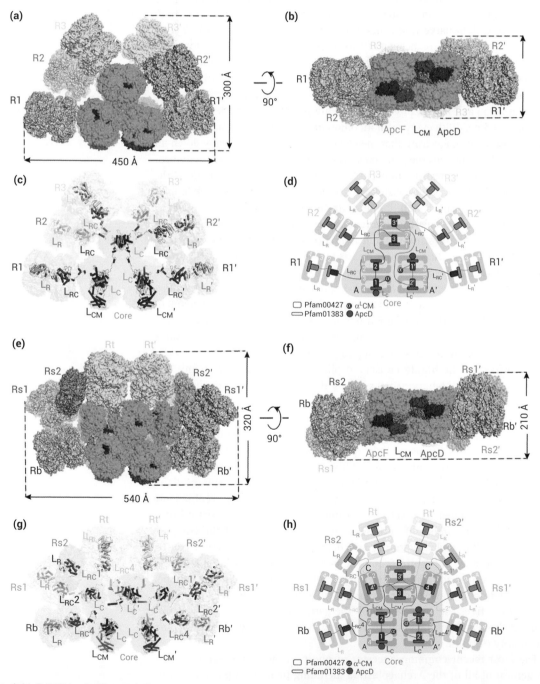

(see Chapter 4), would have contained **phycobilisomes** within its cyano-bacterial endosymbiont. Even today, extant red algae and glaucophytes still contain phycobilisomes that are even more complex than those found in extant cyanobacteria. For example, each of the phycobilisomes of the red alga, *Griffithsia pacifica* is a massive 13.8 megadaltons in size, being made up of no fewer than 862 protein subunits and containing over 2048 pigment molecules. At some point in the early stages of the evolution of photosynthetic eukaryotes there was a divergence in the lineage leading to the Viridiplantae. For reasons that have yet to be discovered, in these organisms phycobilisomes were lost and their function was instead fulfilled by smaller light-harvesting complexes. As shown in Fig 3.7, these LHCs are more fully integrated into the PSI and PSII complexes compared to phycobilisomes.

Cyanobacteria also contain a series of high-light-inducible, single trans-membrane helix proteins, termed Hlips, that bind both chlorophylls and ca-rotenoids. These proteins participate in the protection of the photosynthetic machinery during repair and assembly but may also have other functions. Cyanobacterial Hlips resemble the progenitors of eukaryotic LHC proteins, which also consist of monomers with one or three transmembrane helices. It is believed that Hlips that were present in the original cyanobacterial endosymbiont of photosynthetic eukaryotes evolved through a series of gene duplications that resulted in a new family of proteins, namely the LHCs. The LHCs of algae and plants are made of three transmembrane helices that bind several chlorophyll and carotenoid molecules with both light-harvesting and photoprotective roles. The LHCs have a characteristic tri-meric organization that is shared with CP26, CP29, and CP24, and within thylakoid membranes they become assembled into large supramolecular complexes as shown in Fig 3.7.

3.4   Photosystem II: the water-plastoquinone oxidoreductase

Photosystem II is the key pigment protein complex in oxygenic organ-isms, being responsible for both water oxidation and oxygen evolution. In a typical PSII complex, upon light absorption by the light harvesting antenna, excitation energy reaches the reaction centre core within a frac-tion of a nanosecond. This triggers electron transfer and initiates a charge separation process as described in Chapter 2. The photochemical pigments bound by the dimeric core consists of two pairs of chlorophyll molecules followed by a pair of pheophytin molecules. The latter are modified chlo-rophyll molecules that lack a bound Mg atom. These six photochemi-cal pigments in PSII are collectively known as P680, where P stands for 'pigment' and 680 represents the wavelength of light that causes bleach-ing when the core pigments absorb light that triggers charge separation. As shown in Fig 3.8, P680 spans the membrane with the water-splitting, oxygen-evolving apparatus of PSII located on the lumenal side of the thylakoid membrane, with the pheophytin molecules located towards the stromal side of the membrane.

Fig. 3.7 Light harvesting complexes of plants.

(a) The Light Harvesting Complex II, LHCII. It is configured as a trimer and each monomer is shown in grey, orange and blue ribbons. To the left a single monomer is shown, to highlight the arrangement of the pigments. Green sticks are chlorophyll molecules, while red sticks are carotenoid molecules. Middle view is a side view from the membrane spanning domain, and the right view shows a top view. **(b)** The PSII-LHCII supercomplex of the pea plant consisting of two PSII complexes, four LHCII, two CP24, two CP26, and CP29, all arranged in a symmetric configuration. **(c)** The PSI-LHCI supercomplex of the pea plant highlighting its four associated light harvesting subunits Lhca1 to Lhca4. **(d)** Side view of the PSII-LHCII supercomplex. **(e)** Side view of the PSI-LHCI supercomplex.

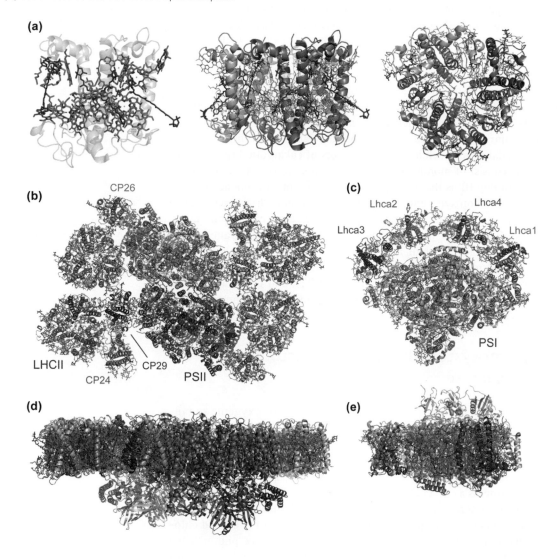

It is thought that Chl_{D1} is the longest wavelength pigment in PSII, which favours the localization of excitation in this particular molecule. Chl_{D1} is the primary electron donor of PSII and initiates charge separation by rapidly transferring its electron to Ph_{D1}, forming the $Chl_{D1}^{+}Ph_{D1}^{-}$ state. Electron transfer then occurs from P_{D1} to Chl_{D1}^{+} and from Ph_{D1}^{-} to the first plastoquinone,

Fig. 3.8 Charge separation in Photosystem II.

The electron transfer cofactors of PSII are shown in sticks. Plastoquinone molecules are shown in orange stick, the non-heme iron (Fe^{2+}) is shown as a dark orange sphere, pheophytin in blue sticks, and chlorophylls in green. The subindex indicates the subunit that provides the main binding to that cofactor, either D1 or D2. The redox active Y_Z-H190 of D1 are shown in grey sticks, but the equivalents in D2 are not shown for clarity. The Mn_4CaO_5 cluster is shown in coloured spheres as indicated. To the left the charge separation sequence is schematized. The process starts with Chl_{D1} reducing Ph_{D1} (1a), followed by rapid oxidation of P_{D1} (1b). Each subsequent step is slower as denoted in the figure. Positive charges will be stored in the cluster, and upon reduction of Q_B this will be exchanged with the quinone pool within the membrane. The non-heme iron is not involved in electron transfer. Y_D-H189, while redox active is only involved in supportive functions and it is not directly involved in water oxidation. The six core pigments of PSII (P_{D1}, P_{D2}. Chl_{D1}, Chl_{D2}, Ph_{D1} and Ph_{D2}) are usually denominated P680.

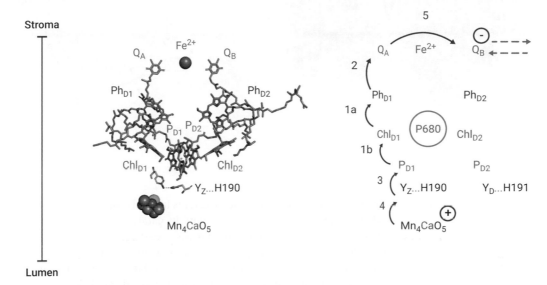

Q_A, forming the $P_{D1}^+Q_A^-$ state. The cation P_{D1}^+ is highly oxidizing and able to generate +1.2 V of electrochemical potential required to oxidize the redox-active tyrosine residue, Y_Z, thereby driving water oxidation. The latter event occurs with a half-time of about 50 nanoseconds. Y_Z provides a hydrogen bond to a nearby histidine, H190. Oxidation of Y_Z leads to proton transfer to H190, producing a neutral radical $Y_Z^\bullet$. This radical then oxidizes the Mn_4CaO_5 cluster, which is the site of water oxidation and oxygen evolution.

The two forms of plastoquinone involved in electron transfer, Q_A and Q_B, are located on the stromal side of the thylakoid membrane. Once it is reduced by an electron from Ph_{D1}^-, Q_A^- transfers its electron to the second and last electron acceptor plastoquinone, Q_B, at a comparatively slow rate of about 400 to 800 microseconds per reaction. Remarkably, although the system began with perfect symmetry, Q_A and Q_B have different properties in PSII. The first acceptor, Q_A, is tightly bound to the protein complex and can only receive and transfer one electron at a time. In contrast, Q_B is mobile and can dissociate from the PSII complex and move freely within the thylakoid bilayer. Q_B can receive two electrons, while taking two protons from the stromal space to form Q_BH_2 or plastoquinol. Once Q_B is doubly reduced, it can leave its binding site on PSII where it is replaced by a freshly oxidized plastoquinone to continue the electron transfer process. The process of quinone exchange occurs with a half-time of about 10 milliseconds.

Fig. 3.9 Photosystem II, the water oxidation cycle, and water channels.
(a) Structure of a dimeric PSII complex indicating the position of the Mn$_4$CaO$_5$ cluster, also known as the Oxygen-Evolving Complex (OEC). **(b)** Water oxidation cycle showing the five basic S states (S$_0$ to S$_4$), the light-induced one-electron oxidation steps (involving Y$_Z$), the proton and oxygen release, and the uptake of the two substrate water molecules. The Mn oxidation states and the reaction times are also given.

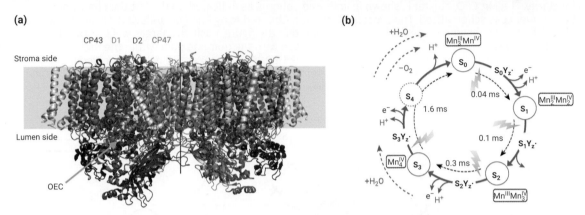

The oxygen-splitting Mn$_4$CaO$_5$ cluster is located on the lumenal face of the thylakoid membrane. This complex and its catalytic function is identical in cyanobacteria, algae, and plants, and has been extensively studied in model organisms from each group. The complex includes a protein-bound metal cluster made up of four manganese ions in oxidation states ranging from +3 to +4 plus a single divalent calcium. The five metals are bridge by O atoms, resulting in the formula Mn$_4$CaO$_5$. In order to oxidize and split two water molecules to 4e$^-$, 4H$^+$, and O$_2$, the Mn$_4$CaO$_5$ cluster must oscillate through five states, termed S$_0$ to S$_4$. S$_0$ is the most reduced ground state featuring three Mn atoms in the oxidation state +3 and single Mn in the +4 state. With every oxidation step the Mn$_4$CaO$_5$ accumulates a positive charge coupled to deprotonation events. In S$_3$ stage, all Mn atoms have reached the +4 state, which means that in the transition to S$_4$, a bridging oxygen rather than Mn is oxidized. It is not until the Mn$_4$CaO$_5$ cluster has reached the S$_4$ step that four electrons are removed from the two substrate water molecules in what appears to be a single step. This results in the regeneration of S$_0$ following the formation of the O-O bond and the release of a dioxygen (O$_2$) molecule (Fig 3.9). A single water-oxidation cycle takes about 2 milliseconds.

3.5   From Photosystem II to Photosystem I: the cytochrome b_6f shuttle

So far, we have followed the light-induced splitting of water and the transfer of the electrons through the PSII complex culminating in the release of the reduced lipophilic electron carrier, plastoquinol. This occurs after four charge separation events, resulting in the formation of two plastoquinol (QH_2) molecules. This process involves taking four electrons from two water molecules, removing four protons from the stromal side of the membrane and transferring another four protons, ultimately originating from water, into the lumenal side of the membrane. The acidification of the lumen generates the proton motive force needed for ATP synthesis as discussed above in section 3.2. Next, the electrons from water splitting must be released from plastoquinone and given sufficient additional reducing power by PSI to enable them to drive metabolic processes such as CO_2 fixation. To get from PSII to PSI, the electrons must travel via the cytochrome b_6f complex, which acts as a shuttle between PSII and PSI.

The cytochrome b_6f complex also uses the power generated during electron transfer to move protons from QH_2, that were originally acquired from the stromal side PSII, into the lumen, while simultaneously picking up two additional protons from the stroma. As noted above, the movement of protons from stroma to lumen creates a strong pH gradient across the thylakoid membrane that can be used to generate ATP via the ATP synthase complex located in the same membrane. Under most conditions, the diffusion of plastoquinone from PSII to the cytochrome b_6f complex via the membrane-located plastoquinone pool, which takes about 10 milliseconds, is the rate-limiting step of oxygenic photosynthesis. In comparison water-oxidation catalysis described in the previous section occurs at least ten times faster. But why are these timings and rates of reaction important? Given that each step of the electron transfer has inherently different rates, it means that the entire flux and stoichiometry of the process needs to be highly balanced and regulated. If conditions are suboptimal in such a way that light cannot be efficiently used for productive photosynthesis, the excess light absorption can result in ROS production, the inhibition of photosynthesis, the breakdown of the photosystems, and even cell damage or death. To prevent this, photoprotective mechanisms have evolved that act over different time ranges and locations in the thylakoid membrane. Some examples of this are described in Case study 3.1.

As shown in Fig 3.10, the cytochrome b_6f complex is a homodimer of eight protein subunits, with the main core subunits having homology to those in the cytochrome bc_1 (complex III) used in mitochondrial respiration. The cytochrome b_6f complex functions via a process known as the Q-cycle, which can be considered in two halves.

In the first half, QH_2 binds to the inner part of the complex on the lumen side where it is oxidized by an iron-sulphur cluster, taking the first electron, and forming the semiplastoquinone radical. This one-electron oxidation event is linked to the release of the two QH_2 protons into the thylakoid lumen. From the iron-sulphur cluster, the electron is then transferred to cytochrome f, which is located on the lumenal side of the complex. Next, the electron is shuttled from cytochrome f to the soluble electron carrier, plastocyanin. After receiving its electron cargo, this small protein is able to dissociate into the lumenal space and eventually transfers the electron to the PSI complex. Meanwhile, the semiplastoquinone radical is oxidized again, but now by a system of three hemes located across the membrane plane (named heme b_n, b_p, and x in Fig. 3.10) which store the second electron.

Case study 3.1
Mechanisms of photoprotection

While light is the main energy source of life on Earth, there is a limit to how much light can be used at a given time. In addition, light intensity is not constant and varies with cloud cover, time of day, and the season. When photosynthesis is working well, all of the light is productively used by the photosystems. However, under stress conditions, for example dehydration or nutrient deficiency, the photosynthetic rate is often reduced. This means that incoming excitation energy from light cannot be used productively (photochemically). The excited chlorophylls can then trigger a change in the electronic configuration of the pigments, entering what is known as the triplet state, which is defined by the position and orientation of the electrons within the orbitals of the chlorophyll rings. This excited triplet state is particularly prone to interacting with molecular oxygen, which upon contact, can cause the formation of highly reactive oxygen species (ROS) that can wreak havoc within the cell.

Other forms of ROS can also be generated at different parts of the photosynthetic machinery including hydrogen peroxide and hydroxyl radicals. To prevent or diminish this, photosynthetic organisms can activate a mechanism known as non-photochemical quenching. This is a way to quench excess excitation energy at the expense of productive photosynthesis. Several ways to achieve quenching have evolved in different organisms to suit their particular light-harvesting apparatus and other needs. For example, in cyanobacteria, there is a soluble carotenoid-binding protein that detects excess excitation. Upon activation it will change conformation allowing it to bind phycobilisomes and efficiently quench the excitation by dissipating it as heat. This protein, known as the Orange Carotenoid Protein (OCP) can also actively scavenge singlet oxygen.

In photosynthetic eukaryotes, non-photochemical quenching is triggered by conformational changes in the light-harvesting complexes of PSII that result in the activation of a dissipative state. These conformational changes are triggered when the cellular pH becomes too acidic and a large difference in pH can be sensed across the thylakoid membrane. This ΔpH activates non-photochemical quenching by protonation of the membrane-bound protein, PsbS, and by triggering enzymatic de-epoxidation of the carotenoid violaxanthin to zeaxanthin, in the PSII light-harvesting complex. The latter process is known as the Xanthophyll Cycle. The exact role of PsbS is unclear but it might mediate the reorganization of LHCII complexes in the thylakoid membrane enabling the Xanthophyll Cycle to occur efficiently. As discussed in Chapter 7, enhancement of photoprotection to improve crop performance is now being investigated.

Fig. 3.10 Structure of the cytochrome b_6f complex and the Q-cycle.

(a) the complex occurs as a dimer in the membrane. Different protein subunits are visualized in different colours. **(b)** detail of cofactor arrangement in a monomer of cytochrome b_6f. **(c)** electron transfer pathways are shown in blue and red arrows, and proton transfer pathways are shown in yellow olive coloured arrows. This structure binds a quinone analogue that occupies the plastoquinone binding side (PQH$_2$). The transmembrane spanning region is marked with a light grey box.

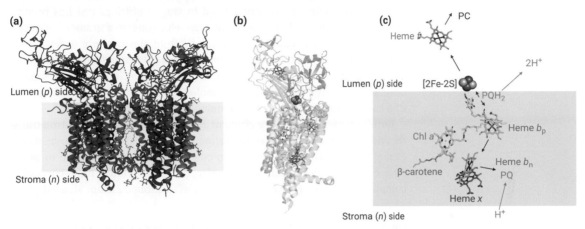

However, it is unclear where exactly in this system the electron is located. After the complete oxidation of the lumenal side plastoquinone this can then exchange for a fully reduced one.

In the second half of the cycle, the first half is repeated with the final result of two electrons now stored in the three-heme system. In the last step, a second but now fully oxidized plastoquinone located in the stromal side of the complex is doubly reduced in a single event by the hemes, taking concomitantly two protons from the stroma and forming the fully reduced QH$_2$, which then exchanges with the membrane. To summarize, the cytochrome b_6f complex transfers electrons from PSII and shuttles them to PSI. During this process some of the energy associated with the electron flux is used to move protons from the stroma to the lumen to create the pH gradient that drives ATP synthesis.

3.6 Photosystem I and ferredoxin reduction to generate NADPH

The third and final part of photosynthetic electron transport involves shuttling the electrons from plastocyanin to ferredoxin, and from ferredoxin to NADPH. The NADPH molecule is a freely mobile electron carrier able to travel to all parts of the cell in order to supply energy in the form of reducing power as required for a host of vital metabolic processes including CO$_2$ fixation and conversion to carbohydrates, lipids, and amino acid biosynthesis, plus defence against ROS, to mention just a few examples. NADPH formation occurs at the stromal side of the thylakoid membrane via a short electron transfer chain involving reduced ferredoxin and the enzyme ferredoxin-NADP$^+$ reductase (FNR). Because PSII is specialized to perform the highly oxidizing chemistry required to split water, its redox cofactors are not fine-tuned to deliver these electrons to ferredoxin and NADPH directly. The latter are highly reducing molecules with midpoint redox potentials of −430 and −320 mV, respectively. Therefore, while PSII evolved to do the oxidizing chemistry related to water splitting, PSI has become specialized to use the light energy to move electrons up a gradient of nearly 0.8 V towards more reducing potentials, ie from plastocyanin to ferredoxin (Fig 3.11). In order to achieve this, the cofactor chlorophylls involved in PSI charge separation are fine-tuned in such a way that, upon charge separation, a chlorophyll at a position homologous to that of Chl$_{D1}$ and Chl$_{D2}$ in PSII can produce over −1.0 V of reductive power.

Both PSII and PSI are very efficient at converting light into electron transfer and PSI is well known for operating at nearly 100% quantum yield efficiency under optimal conditions. In contrast, PSII works at about or below 90% efficiency depending on the size of its associated light-harvesting complexes and on the particular species of cyanobacterium, alga, or plant.

Fig. 3.11 Structure of PSI bound to ferredoxin and plastocyanin.

A membrane plane view of the complete triple complex. Pc denotes plastocyanine and Fd ferredoxin. The redox cofactors are highlighted as sticks and spheres, while the protein scaffold and antenna pigments of PSI are shown in transparency. Q_A and Q_B denote the phylloquinones. The Cu$^+$ of Pc, F_X, F_A, F_B of PSI and the Fe$_2$S$_2$ cluste rof Fd are shown as spheres.

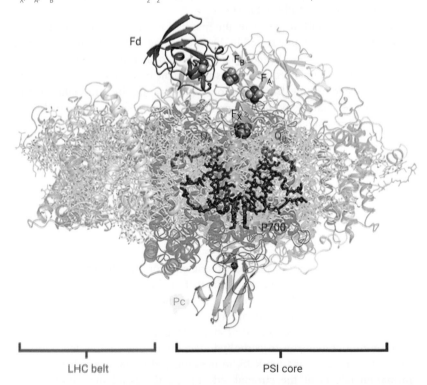

This still means that almost every photon absorbed by a pigment in PSI will result in the initiation of charge separation at the reaction centre. In PSII, electron transfer initiates within the D1 side, reaching Q_A that is bound by D2, then moving to Q_B bound by D1. In consequence, there is no charge separation occurring from the D2 side or going from Q_B to Q_A. Remember that the photosystem has a symmetric configuration. However, in PSI charge, separation can occur on either side of the reaction centre, although not on both sides at the same time. Thus, even though PSI is a heterodimer, it still retains a high degree of 'homodimeric' function. There also seems to be a degree of asymmetry in charge separation as it is thought that the PsaA dominates over the PsaB side given that electron transfer goes via the A side for about 60% of the events.

As we saw above, in PSII the chlorophyll cation is localized and stabilized in P_{D1}. This is different in PSI where the chlorophyll cation is shared between the pair of chlorophyll pigments equivalent to P_{D1} and P_{D2} of PSII. This pair of chlorophylls in PSI is usually called P700. A further difference between the two photosystems is found in their respective charge separation processes. Here, in contrast to PSII, PSI has strongly bound, non-exchangeable quinones, that serve as single-electron shuttles between the photochemical chlorophylls and the iron-sulphur (Fe_4S_4) acceptor protein cluster, F_X. The nature of the quinones in PSII and PSI is also different, with the latter using phylloquinone instead of plastoquinone. The reason for this difference is that plastoquinone operates at a more oxidizing potential, which is suitable for its job as part of the water oxidation process. In contrast, phylloquinone operates at a more reducing potential, as appropriate to facilitate electron transfer to the Fe_4S_4 cluster.

After the initiation of charge separation in PSI, the primary acceptor, F_X, transfers its highly reducing electrons through two additional iron-sulphur clusters, and from there to ferredoxin, which is a small soluble iron-sulphur (Fe_2S_2) protein that is able to attach to the stromal side of PSI. Once it is reduced by electrons from PSI, ferredoxin is in turn oxidized by ferredoxin-NADP$^+$ reductase (FNR), to generate NADPH. FNR is a soluble oxidoreductase that can be free in the stroma or bound to the thylakoid membrane in the vicinity of PSI as shown in Fig 3.3. Ferredoxin can also dissociate from the thylakoid membrane and is able to function independently as a reducing agent in several important energy-requiring metabolic processes including nitrogen fixation and assimilation, amino acid synthesis, and sulphur metabolism.

The process of electron transport from water to NADPH described above is usually referred to as 'linear electron flow' and it involves ATP and NADPH being formed at a constant ratio. However, depending on the cell's needs, ATP and NADPH may be required in different amounts. To achieve a more modulated balance of ATP and NADPH production, a process known as cyclic electron flow (CEF) evolved before the most recent common ancestor of cyanobacteria. Linear electron flow (LEF) from water to NADPH typically produces an ATP to NADPH ratio of about 1.3 : 1, but this falls short of the 1.5 : 1 ratio required to support CO_2 fixation. To achieve an increased ratio of ATP to NADPH, both cyanobacteria and chloroplasts use the CEF pathway, which produces ATP independently of NADPH. In CEF, electrons are cycled through the electron transfer chain, resulting in additional proton pumping without the electrons reaching NADPH. After PSI reduces ferredoxin the latter reduces a large thylakoid membrane protein complex, known as NDH-1, which is homologous to Complex I in the respiratory electron transfer chain of mitochondria. NDH-1 stands for NADH dehydrogenase-like complex type-1. However, in photosynthesis, this complex uses ferredoxin and not NADH as its electron donor. After taking electrons from ferredoxin, NDH-1 reduces plastoquinone and uses the flow of electrons to pump protons across the membrane. Reduced plastoquinol can then return to the cytochrome b_6f complex, closing the cycle with the re-reduction of plastocyanin.

3.7   Photosynthetic carbon metabolism

Light-driven CO_2 fixation and carbon assimilation are key activities of photosynthetic autotrophs, from cyanobacteria to plants.

They involve the conversion of inorganic substances, such as CO_2, nitrogen compounds, and minerals into complex organic molecules, such as carbohydrates, lipids, proteins, and poly-nucleic acids. These anabolic reactions require NADPH and ATP, which, as discussed above, are produced during the light reactions of photosynthesis.

Rubisco and CO_2 fixation

In most photosynthetic organisms, CO_2 is initially fixed via conversion into simple sugar units in the **reductive pentose phosphate** (RPP) pathway. The first step in this process involves the combination of a CO_2 molecule with the pentose sugar, ribulose bisphosphate, leading to the formation of two molecules of 3-phosphogylcerate as summarized in Fig 3.12. For each three turns of the cycle, three CO_2 molecules are fixed and one 3-carbon triose phosphate is produced in a highly energy demanding process that uses 6 NADPH and 9 ATP molecules. As described above, the latter are produced during photosynthesis and CO_2 fixation is normally closely coupled with the operation of the light reactions.

The enzyme that catalyses the initial CO_2 fixation reaction is **rubisco** (RibUlose BISphosphate Carboxylase/Oxygenase). Rubisco is not unique to photosynthetic organisms and is found in all major groups of life, including archaea. The importance of rubisco in land plants is shown by its abundance; it makes up 30–50% of the total soluble protein of leaves and it is estimated that there are 5 kg of rubisco for every person on the Earth. Rubisco in plants and algae is a large enzyme of mass ≈550 kDa. The complete protein consists of 8 copies of a large (L) ≈50 kDa subunit, plus 8 copies of a small (S) ≈15 kDa subunit. These monomers are arranged in a 16-subunit L_8S_8 configuration as shown in Fig 3.13. One of the most striking things about rubisco is that, despite its vital role in photosynthesis, it is inefficient as a biological catalyst. Although the main catalytic reaction of rubisco is carboxylation, it has an alternative oxygenation reaction that only generates half of the yield of the 3-phosphogylcerate product compared to the main reaction (see Fig 3.15). This wasteful side reaction occurs about one-third of the time and greatly reduces the overall catalytic efficiency. Rubisco

Fig. 3.12 The Reductive Pentose Phosphate (RPP) pathway.

The RPP pathway is the metabolic route whereby photosynthetically fixed CO_2 is converted to sugar phosphates. The overall reactions in three turns of the cycle involve the fixation of three molecules of CO_2 to produce one molecule of the C3 sugar phosphate, glyceraldehyde 3-phosphate (GAP) for export for conversion into more complex sugars. The three turns of the cycle require three molecules of NADPH to provide the reductant, plus nine molecules of ATP to drive this highly energy demanding process. Meanwhile, the other five molecules of glyceraldehyde 3-phosphate are recycled to regenerate the C5 ribulose bisphosphate substrate for further CO_2 fixation.

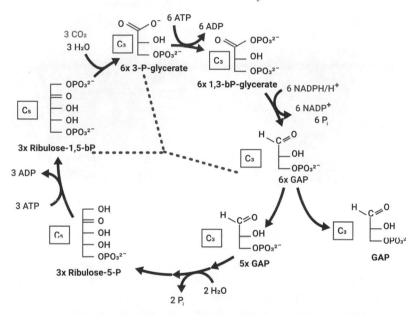

is also a very sluggish enzyme with a turnover frequency typically between 1 and 10 s^{-1} while most other enzymes have more rapid rates of 10^4 to 10^6 s^{-1}. Part of the reason for the inefficiency of rubisco is that it originally evolved in a highly anaerobic environment that persisted for as long as 1.5 billion years. This meant that the competing oxygenation reaction did not become problematic until photosynthetic organisms had become locked into a dependence on rubisco.

It is likely that the original form of rubisco was an L_8 oligomer that was probably more catalytically efficient, but also much more oxygen-sensitive than the present form. After GOE1 at about 2.4 Ga, increasing oxygen levels meant that the unwanted oxygenase side reaction became more prevalent. In order to reduce the frequency of the side reaction, the ancestral rubisco oligomer evolved into a slower but less oxygen-sensitive L_8S_8 configuration. The trade-off was that the new form of rubisco sacrificed some of its catalytic efficiency in order to function in the new high oxygen environment. A hypothetical scheme for the evolution of rubisco is shown in Fig 3.14. Here rubisco originates from an enolase-like enzyme that had emerged prior to photosynthesis and was probably already present in LUCA. This ancestral enzyme had initially evolved in a heterotrophic context, possibly as part of nucleotide metabolism, before it eventually assumed its current role in autotrophic carbon metabolism. Eventually the

Fig. 3.13 Structure of rubisco in plants.

This shows the structure of the hexadecameric Form I of rubisco found in plants and algae, in this case in spinach. The evolution of Form I rubisco from its earlier precursors is shown in Fig 3.13. Here, large rubisco subunits, RbcL, are shown alternating in white and grey and small subunits, RbcS, in yellow. There is a single antiparallel RbcL dimer (RbcL in green and RbcL' in blue) and the adjacent RbcS subunits in the side-view are depicted in a ribbon representation. The antiparallel RbcL dimer is shown with the RuBP substrate molecules (in red) bound to the active sites.

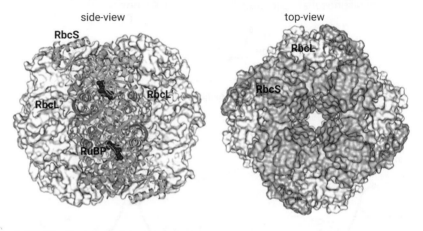

side-view   top-view

© 2017 The Protein Society

enzyme assumed its present L_8S_8 configuration in cyanobacteria and this is the form of rubisco that was passed on to all photosynthetic eukaryotes following the primary endosymbiotic event about 2.2 Ga.

In extant cyanobacteria, the efficiency of rubisco has been enhanced by sequestering it into subcellular compartments termed carboxysomes. Although photosynthetic eukaryotes do not contain carboxysomes, they do employ chaperonin proteins that are required for the assembly of functional rubisco oligomers. The eukaryotic rubisco-specific chaperonin, Cpn60, is homologous to a bacterial protein, GroEL, meaning that both proteins probably evolved prior to 2.2 Ga. One set of problems faced by rubisco is the frequent occurrence of catalytic errors and the accumulation of inhibitors, particularly under low-light conditions. A second class of chaperonins, called rubisco activases, play important roles in the metabolic repair of damaged rubisco oligomers. The discovery of these various classes of chaperonins was important because biotechnological strategies to improve the functionality of rubisco will also require the participation of the appropriate chaperonins in order to generate and maintain the active form of the enzyme (see Chapter 7).

Photorespiration

Because rubisco can function as both a carboxylase and an oxygenase it can either react with CO_2 to generate two molecules of 3-phosphoglycerate, or with O_2 to generate two molecules of 3-phosphoglycerate plus a two-carbon compound known as 2-phosphoglycolate (see Fig 3.15). This

Fig. 3.14 Hypothetical timeline for the evolution of rubisco.

The figure summarises the individual events during the emergence of rubisco from a non-CO_2-fixing ancestor to the more complex contemporary rubisco (Form I), which operates in chloroplasts and cyanobacterial carboxysomes. The precursor rubisco-like protein (RLP) was possibly present in LUCA before 4 Ga and initially evolved in a CO_2-rich, anoxic environment. The relatively simple RLP then evolved into heterotrophic (Form III) and autotrophic (Form II) complexes of progressively higher stoichiometries. Ultimately the enzyme became associated with a chaperone and acquired the 16-subunit, L_8S_8 configuration (Form I) that is now found in extant cyanobacteria and chloroplasts.

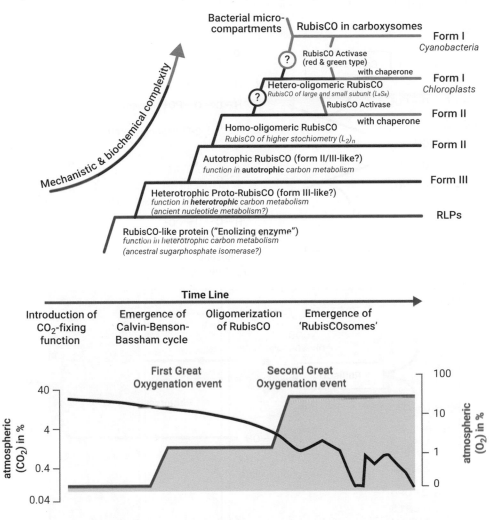

oxygenase reaction is considered wasteful because the 2-phosphoglycolate product must then be recycled via peroxisomes and mitochondria in order to recover the more useful product, 3-phosphoglycerate. Depending on conditions the oxygenase side reaction can require as much as 50% of additional energy expenditure compared to carboxylation. This wasteful reaction occurs because the enzyme is unable to differentiate between the small gaseous molecules CO_2 and O_2. Also, while CO_2 is a trace gas measured in few hundred parts per million, O_2 makes up 21% of the atmosphere.

Fig. 3.15 Photorespiration and rubisco.

(a) The main reaction of rubisco in the presence of CO_2 is the carboxylation of its substrate to yield two molecules of 3-phosphglycerate (upper reaction). However, in the presence of oxygen, rubisco also catalyses an oxygenation of its substrate that only yields one molecule of 3-phosphglycerate, plus an unwanted side product, 2-phosphglycerate (lower reaction). The oxygenation reaction significantly reduces the efficiency of CO_2 fixation and hence that of overall photosynthetic productivity. **(b)** Photorespiration is the consequence of the rubisco oxygenase side-reaction and its unwanted product, 2-phosphoglycolate, must be cycled through a complex series of reactions in peroxisomes and mitochondria in order to generate 3-phosphoglycerate.

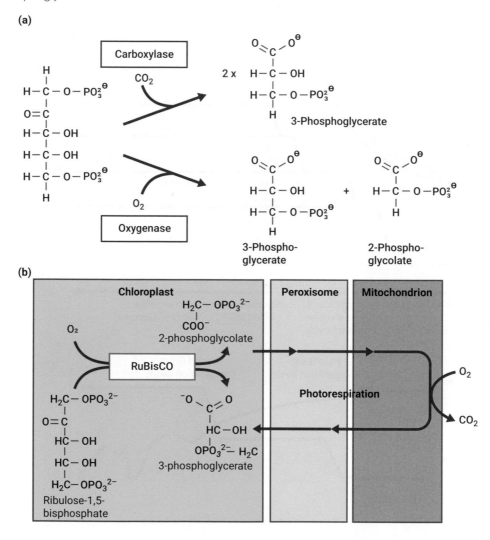

Photorespiration is defined as the rubisco oxygenation reaction plus the subsequent metabolism required to produce assimilable intermediates or to regenerate 3-phosphoglycerate.

In plants photorespiration occurs in chloroplasts, peroxisomes, and mitochondria. Briefly, 2-phosphoglycolate formed in chloroplasts is dephosphorylated to glycolate, and transported to peroxisomes for deoxygenation to form glycine. This generates hydrogen peroxide, which is detoxified by catalase in peroxisomes. In mitochondria, glycine is converted to serine, which travels back to peroxisomes to be transformed into glycerate. Glycerate is returned to the chloroplast where it is phosphorylated to form 3-phosphoglycerate to re-enter the RPP pathway.

Photorespiration is an ancient process that likely predates cyanobacteria. Both cyanobacteria and photosynthetic eukaryotes have evolved multiple mechanisms to avoid or minimize the rubisco oxygenation reaction. A strategy not inherited by eukaryotes is the use of carboxysomes as discussed in the previous section. In many algae and liverworts, a subcellular microcompartment known as the pyrenoid has evolved to concentrate CO_2. This microcompartment is made up of mostly rubisco and unlike the carboxysome it does not have a shell. Instead, rubisco forms a matrix that behaves as a phase-separated liquid-like organelle. Another way to avoid photorespiration is to increase CO_2 concentrations in the vicinity of rubisco molecules. In angiosperms several carbon concentration mechanisms, known as the C4 and CAM pathways, have evolved more recently in an attempt to bypass or at least minimize photorespiration.

Carbon assimilation via the C4 and CAM pathways

The conventional form of CO_2 fixation and assimilation into carbohydrates is C3 photosynthesis, since the initial product is the 3-carbon compound, 3-phosphogylcerate. Most plants use this pathway alone, but the process becomes less efficient as the ratio of CO_2 to O_2 decreases. About 35 Ma, numerous angiosperm groups, including many grass species, independently developed a CO_2-concentrating mechanism known as C4 photosynthesis that enabled them to maximize the intracellular CO_2 concentration close to the rubisco active site. This mechanism did not become widespread until about 6 Ma when a combination of lower atmospheric CO_2 levels, higher temperatures, and increasing aridity favoured the spread of C4 photosynthesis across the globe. C4 photosynthesis has evolved independently at least 62 times in 18 plant families. It is particularly useful at higher temperatures where the ratio of CO_2 to O_2 in solution is markedly decreased. Because it also results in increased water-use efficiency, C4 photosynthesis is highly advantageous in arid or semi-arid conditions. The increased efficiency of this mechanism is shown by the fact that, although C4 plants only make up about 5% of land plant biomass and 3% of plant species, they account for about 23% of terrestrial CO_2 fixation.

The pathways of C4 and CAM (Crassulacean Acid Metabolism) photosynthesis are shown in Fig 3.16. Similar to several other aspects of land plant evolution discussed in Chapter 5, a number of 'pre-adaptations' for the development of C4 photosynthesis seem to have already been present in some plant groups. For example, C4 photosynthesis is most frequently found in plants that already had features such as extensive vascular bundle sheath tissue that enabled the spatial separation of the two parts of the process. In most cases, enzymes and pathways that were not involved in photosynthesis supplied the components required for the assembly of a full C4 pathway

Fig. 3.16 Summary of the C3, C4 and CAM carbon assimilation pathways.

Plants have at least three alternative mechanisms to assimilate photosynthetically fixed CO_2. The most common is the C3 pathway where CO_2 is directly fixed by rubisco. The C4 pathway is an adaptation to lower atmospheric CO_2 levels, higher temperatures, and increasing aridity. CO_2 is initially fixed into 4-carbon compounds via PEP carboxylase in mesophyll cells before being released in concentrated form close to the rubisco active site in bundle sheath cells. The CAM pathway resembles the C4 pathway, however, the C4 production step occurs at night, and the decarboxylation and rubisco fixation occurs during the day.

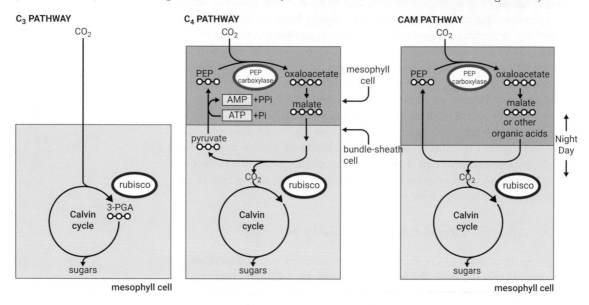

and partial versions of this pathway are also found in C3 plants. The recruitment of such enzymes for a new stand-alone C4 pathway was facilitated by genome duplications that generated 'spare' copies of the relevant genes and enzymes. The pathway can be summarized thus. In mesophyll cells, CO_2 is hydrated by carbonic anhydrase to generate bicarbonate, which is used by phosphoenolpyruvate carboxylase to generate oxaloacetate, a 4-carbon compound. Oxaloacetate is converted to malate or aspartate, which then diffuses through **plasmodesmata** into bundle sheath cells where it is decarboxylated to pyruvate, releasing the CO_2 in the vicinity of rubisco thereby re-entering the RPP cycle (Fig 3.15). The advantage of this process is that carbonic anhydrase has an affinity for CO_2 several orders of magnitude greater than rubisco. This allows C4 plants to increase the CO_2 concentration around rubisco by 15-fold.

The oldest groups of C4 plants, originating between 35 to 20 Ma, contain some of today's most productive crops, such as sugar cane, maize, and sorghum. The most recently emerged clades are endemic plants, that might be just a few hundred thousand years old, such as several *Flaveria* species that are rare Mexican relatives of the sunflowers. The competitive advantage of C4 plants over C3 plants is clear in hot climates, and under conditions of drought and nitrogen limitation. The temperature dependency relates to the fact that the rubisco oxygenation reaction becomes increasingly prevalent at temperatures above 20°C at today's atmospheric CO_2 concentrations.

However, C4 carries a major cost: the additional ATP needed to drive CO_2 concentration lowers its photosynthetic efficiency relative to C3 particularly under conditions where photorespiration is naturally suppressed, such as high CO_2 and cool temperatures.

Crassulacean Acid Metabolism (CAM) is an alternative form of C4 photosynthesis which some plants have evolved. In CAM, the carbon fixation and assimilation reactions occur in the same cell, but are separated temporally. CAM is found in many drought-tolerant species where there is an overriding need to conserve water, and especially to prevent its loss through stomata during the day. During the night, plants keep their stomata open, exchanging gases and routing CO_2 through the C4 assimilation pathway from where it is stored as malic acid in the vacuole. During the day, with the stomata closed, the stored malic acid is released and decarboxylated in the chloroplast for fixation by rubisco. Like C4 photosynthesis, CAM has evolved independently many times. Although mainly found in flowering plants CAM is also present in some lycopods, ferns and gymnosperms. Within the angiosperms, CAM is most commonly found in succulents and bromeliads.

3.8   Nitrogen assimilation

Nitrogen assimilation is second only to that of carbon in its importance for the metabolism of photosynthetic organisms.

While CO_2 fixation produces carbohydrates, inorganic nitrogen is required for their further metabolism to generate amino acids and nucleic acids and thousands of other organic nitrogenous compounds. Inorganic nitrogenous compounds, such as nitrates and ammonia, are ultimately derived from N_2 gas, which is a highly abundant molecule that makes up 78% of the atmosphere. Cyanobacteria, along with many other bacteria, are able to fix atmospheric N_2 via an enzymatic complex-celled nitrogenase. This enables them to directly generate their own organic nitrogenous compounds from N_2 gas. However, eukaryotic photosynthetic organisms lack this ability and must use non-gaseous inorganic nitrogenous compounds, such as nitrates or ammonia that are already present in the environment, eg in the soil. Alternatively, some plants such as legumes, are able to form symbioses with N_2-fixing bacteria thereby reducing their dependence on soil nitrates. As discussed in Chapter 7, one of the goals for future crop improvement is to use biotechnology to transfer the capacity for N_2-fixation into major crop groups such as cereals.

In most plants and algae the main source of nitrogen is nitrates, which are present as soluble minerals in the soil or water. As shown in Fig 3.17, nitrate reduction is a two-stage process that yields ammonium (NH_4^+) as its end product. The first step is catalysed by nitrate reductase, a large, complex homodimer that is highly regulated at both the enzyme and gene levels. Nitrate reductase activity is determined by its phosphorylation status, which is itself affected by factors such as calcium, O_2 and CO_2 concentration, photosynthetic rate, and okadaic acid. Transcription of the nitrate

Fig. 3.17 Nitrogen-assimilation pathway in higher plants.

Nitrates are taken up from the soil or water by plants and algae via plasma membrane carrier proteins and initially reduced by a cytosolic nitrate reductase. Further metabolism to nitrites, ammonium, and amino acids occurs in plastids. The specific steps shown include: nitrate transporters (NRT), nitrate reductase (NR), nitrite reductase (NiR), ammonium transporters (AMT), glutamine synthase (GS), glutamate synthase (GOGAT), asparagine synthase (AS), glutamate dehydrogenase (GDH), and isocitrate dehydrogenase (ICDH).

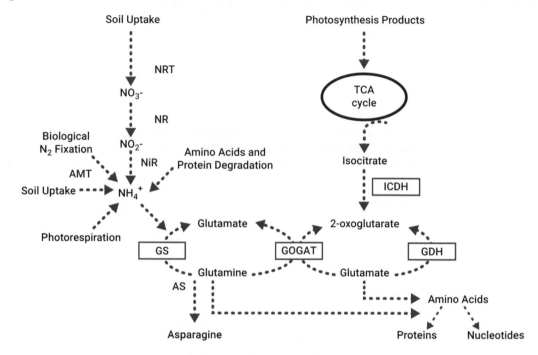

reductase gene is regulated by factors such as light, circadian rhythms, sucrose, cytokinin, and glutamine. The high degree of control exerted over nitrate reductase is necessary because its product, nitrite, is potentially toxic and is normally kept below 15 nmol.g^{-1} fresh weight in plant leaves. The second reduction step is catalysed by nitrite reductase, a ferredoxin-dependent enzyme that reduces nitrite ions to NH_4^+ ions.

Inorganic ammonium is converted into amino acids by glutamine synthase and glutamate synthase, which effectively add an amino group to 2-oxoglutarate to produce glutamate. Glutamine synthase is an octomeric homodimer with cytosolic and plastidial isoforms regulated by a complex set of factors including light, nutrient availability, metabolites, and possibly by phosphorylation. Glutamate synthase is generally found as two plastidial monomeric isoforms that respectively use ferredoxin and NADH as reductants. While not as highly regulated as glutamine synthase, its gene expression and enzyme activity are still subject to a wide range of effectors including light and sucrose. Other amino acids, such as aspartate and asparagine, can be formed from glutamate via aminotransferase reactions. Amino acids in turn are precursors for a huge range of other important compounds including nucleic acids, proteins, porphyrins, and lignins.

 Chapter summary

- Oxygenic photosynthesis occurs on the thylakoid membrane system, as found in cyanobacteria, algae, and plants.
- The first light reaction occurs on PSII where water is split and the resulting electrons are transferred via the cytochrome b_6f complex to PSI. Water splitting is catalysed by the Mn_4CaO_5 cluster which transitions through nine steps before formation of molecular oxygen from two water molecules.
- A second light reaction occurs on PSI to generate electron carriers like ferredoxin and NADPH that are able to power the reductive reactions of cell metabolism. During this electron transfer, protons are pumped across the thylakoid membrane to generate a pH gradient able to power ATP synthesis.
- Thylakoid membranes and their lipid and protein components are highly dynamic and are able to adjust rapidly to different environmental conditions.
- CO_2 fixation by rubisco is a vital reaction but is relatively inefficient due to a competing oxygenase reaction. In plants, carbon concentrating mechanisms, such as C4 and CAM photosynthesis have recently evolved to deal with this.
- Eukaryotes cannot fix nitrogen but can either obtain nitrates from the soil or water or form symbioses with N_2-fixing bacteria.

Further reading

Albanese P, Tamara S, Saracco G et al (2020) How paired PSII–LHCII super-complexes mediate the stacking of plant thylakoid membranes unveiled by structural mass-spectrometry. *Nature Commun* 11, 1361. DOI: https://doi.org/10.1038/s41467-020-15184-1
Describes the molecular mechanism that help maintain grana stacking in plants.

Bhaduri S et al (2020) A novel chloroplast super-complex consisting of the ATP synthase and photosystem I reaction center. *PLOS ONE* 15, e0237569. https://journals.plos.org/plosone/article?id=10.1371/journal.pone.0237569
The discovery and characterization of a novel supercomplex in spinach formed by PSI, ATP synthase, and FNR.

Croce R, van Amerongen H (2020) Light harvesting in oxygenic photosynthesis: Structural biology meets spectroscopy. *Science* 369, 933–942. DOI: 10.1126/science.aay2058. https://science.sciencemag.org/content/369/6506/eaay2058?rss%253D1=
Recent review of the structure and function of photosynthetic supercomplexes.

Erb TJ, Zarzycki J (2018) A short history of RubisCO: The rise and fall (?) of nature's predominant CO_2 fixing enzyme. *Curr Opin Biotechnol* 49, 100–107. DOI: http://dx.doi.org/10.1016/j.copbio.2017.07.017
A hypothesis for the non-photosynthetic origins and subsequent evolution of rubisco.

Giovagnetti V, Ruban AV (2018) The evolution of the photoprotective antenna proteins in oxygenic photosynthetic eukaryotes. *Biochem Soc Trans* 46, 1263–1277. DOI: 10.1042/BST20170304

Photoprotective mechanisms in photosynthetic eukaryotes.

Ibrahim M et al (2020) Untangling the sequence of events during the $S_2 \rightarrow S_3$ transition in photosystem II and implications for the water oxidation mechanism. *Proc Nat Acad Sci* 117, 12624–12635. https://pubmed.ncbi.nlm.nih.gov/32434915/

Detailed structural analysis of the water splitting process.

Kirchhoff H (2019) Chloroplast ultrastructure in plants. *New Phytol* 223, 565–574. DOI: https://doi.org/10.1111/nph.15730

A comprehensive review of the current understanding of the higher order organization of the thylakoid membranes.

Mares, J et al (2019) Evolutionary patterns of thylakoid architecture in cyanobacteria. *Front Microbiol* 10, 277. DOI: 10.3389/fmicb.2019.00277

Detailed study of the diversity of thylakoid membrane architectures in cyanobacteria.

Niklaus M, Kelly S (2019) The molecular evolution of C4 photosynthesis: Opportunities for understanding and improving the world's most productive plants. *J Exp Bot* 70, 795–804. DOI: 10.1093/jxb/ery416

A recent review on the diversity and evolution of C4 plants.

Strašková A et al (2019) Pigment-protein complexes are organized into stable microdomains in cyanobacterial thylakoids. *Biochim Biophys Acta Bioenerg* 1860, 148053. DOI: https://doi.org/10.1016/j.bbabio.2019.07.008

Organization of the photosynthetic apparatus in cyanobacteria thylakoid membranes.

 Discussion questions

3.1 Why do photosystems require the association of additional light-harvesting complexes?

3.2 Why is the cytochrome b_6f complex necessary in oxygenic photosynthesis?

3.3 Beyond water oxidation, what are other functional differences between PSII and PSI?

3.4 What is the purpose of carbon concentrating mechanisms?

4 ENDOSYMBIOSIS: HOW EUKARYOTES ACQUIRED PHOTOSYNTHESIS

Learning objectives

- Introducing the cyanobacterial partners whose descendants still carry out all of the photosynthesis in eukaryotes.

- Describing how a particular eukaryotic cell was able to engulf a cyanobacterium and retain it intact instead of digesting it.

- Examining how endosymbiosis enabled a cyanobacterium to evolve into plastid organelles inside their host cell.

- Surveying the ways in which plastids have diversified into several different multifunctional organelles in eukaryotes.

- Understanding the wider evolutionary context of these processes in terms of their timing and global significance.

4.1 Introduction

In Chapter 2, we saw that during the early-to-mid Archean Eon, between 3 and 4 Ga, several different forms of photosynthesis evolved in various groups of bacteria. Most of these photosynthetic mechanisms were anoxygenic and unable to use water as an electron source. However, a more efficient mechanism also emerged, namely oxygenic photosynthesis. This enabled one group of chlorophyll a-containing prokaryotes called the cyanobacteria to use highly abundant liquid water, as a hydrogen donor with the release of molecular oxygen as a by-product. In contrast, anoxygenic photosynthesis relies on less abundant hydrogen donors, such as hydrogen gas (H_2), hydrogen sulphide (H_2S), or sulphur (S_8). This meant that the oxygenic cyanobacteria tended to have much higher growth rates than the various groups of anoxygenic photosynthetic bacteria. Although it might have originally been present in other groups of bacteria during the Archean, by the Proterozoic Eon (after 2.5 Ga), at this time oxygenic photosynthesis was confined to the cyanobacteria. Over the next billion years the cyanobacteria became increasingly dominant organisms across the world in terms of their biomass and widespread distribution.

In this chapter, we will look at how one of these oxygenic cyanobacterial cells was probably taken up by a scavenging heterotrophic eukaryotic cell. But, instead of immediately digesting the bacterial cell as would occur normally, the eukaryotic host appears to have retained it intact, eventually converting it into a completely new type of organelle called the plastid. This remarkable feat was achieved via the process of endosymbiosis, and it gave rise to a radically different form of biological life. The resulting hybrid organism was an oxygenic, eukaryotic autotroph that was the ultimate ancestor of all of the algae and land plants. These organisms eventually became the major primary producers in the biosphere and still dominate it today in terms of their total biomass. The eukaryotic group resulting from that unique primary endosymbiotic event over two billion years ago is called the Archaeplastida. As discussed in Chapter 1, the subsequent radiation of these and other plastid-containing organisms, has fundamentally changed both the biology and chemistry of planet Earth. For example, the acquisition of oxygenic photosynthesis by eukaryotes contributed significantly to the transformation of the atmosphere and oceans from being largely anaerobic habitats, only able to support simple life forms, to the oxygen-rich aquatic and terrestrial ecosystems that sustain the vast majority of life today. The broader significance of endosymbiosis in evolution is discussed in Case study 4.1.

Case study 4.1
The significance of endosymbiosis in evolution

Endosymbiosis is a form of symbiosis where one partner, the endosymbiont, is fully engulfed by and resides within another partner, the host. In some cases the endosymbiont maintains its autonomy and can survive outside the host cell. More commonly, the endosymbiont undergoes selective gene loss and eventually loses its autonomy. Endosymbiosis has played a crucial role in the evolution of eukaryotes and there is good evidence that two of their most important organelles, mitochondria and plastids, are derived from ancient endosymbiotic events.

Mitochondria are probably the much-reduced descendants of an aerobic α-proteobacterial cell that was taken up by a host cell that was an archaeon. Once they were fully incorporated into their host cell, mitochondria served as highly efficient providers of energy via respiration. The resulting combined cell possibly evolved into an aerobic eukaryote capable of more rapid growth and much greater size than prokaryotes.

Plastids are descendants of a cyanobacterium that was taken up by a eukaryote that already contained a mitochondrial endosymbiont. The incorporation of a cyanobacterium endosymbiont is evidently very rare, with almost all photosynthetic eukaryotes being descended from a single event that probably occurred about 2.0 Ga. The result was a eukaryotic cell capable of aerobic respiration that was also able to carry out oxygenic photosynthesis.

The two primary endosymbioses that led to mitochondria and plastids were extremely rare, but momentous, evolutionary events that produced radically new lineages of life, namely the eukaryotes and the non-prokaryotic photosynthesizers. Subsequently there have been numerous instances of secondary and **tertiary endosymbiosis** whereby eukaryotic cells have taken up entire algae (rather than cyanobacteria) and converted them into photosynthetic organelles. The fact that it is much more common means that this process is easier than primary endosymbiosis of a cyanobacterium. These forms of endosymbiosis have generated many new groups of photosynthetic eukaryotes, including diatoms, dinoflagellates, and brown algae, that now play key roles in the major marine ecosystems.

4.2   The cyanobacterial partner

The acquisition of photosynthesis by eukaryotes resulted from an endosymbiotic association between a cyanobacterium and a heterotrophic eukaryotic cell that already contained mitochondria and was capable of aerobic respiration.

The origins of the cyanobacteria can be traced back to their most recent common ancestor that was most likely already present early in the Proterozoic Eon at about 2.2 to 2.4 Ga, and possibly much earlier. The cyanobacterial ancestor was a relatively complex oxygenic phototroph with two photosystems that were structurally almost indistinguishable from those that we see today in plants and algae. Because these two photosystems are so highly conserved, their core structures must have evolved very slowly with very few adaptive changes over the subsequent two billion years since the initial endosymbiotic event. Indeed, as shown in Fig 4.1, the ultrastructure of a typical cyanobacterial cell still has many similarities to its chloroplast descendants today, and the same is true for their fundamental photosynthetic molecular machinery. This implies that the original cyanobacterial photosystems, and their unique thylakoid glycolipids, were already very well adapted for their major functions as biological machines capable of light-driven water-splitting and electron transport for the efficient production of ATP and NADPH.

The impressive stability of the photosynthetic apparatus over time has implications for current efforts to bioengineer more efficient versions

Fig. 4.1 Comparison of cyanobacterial cell with a typical plant chloroplast.

(a) Ultrastructure of the cyanobacterium, Synechocystis, showing various subcellular components. L, Lipid droplet; C, Carboxysome; PHB, Polyhydroxybutyrate granule; PP, Polyphosphate body; Glyc, Glycogen granule; Cyano, Cyanophycin granule. **(b)** Generalised Structural similarities between cyanobacteria and chloroplasts.

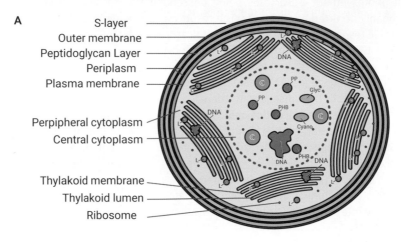

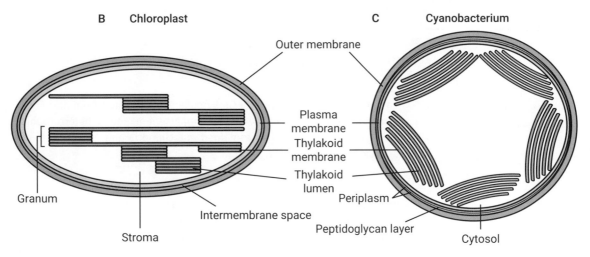

for improved food production and climatic resilience (see Chapter 7). We know from fossil and molecular evidence that, during the last 2 to 3 billion years, oxygenic organisms have undergone many evolutionary changes. For example, they evolved from tiny unicellular aquatic bacteria and algae into some of the largest and longest living terrestrial organisms on Earth. Today's photosynthetic organisms range from microscopic phytoplankton in the lakes and oceans to enormous multicellular land plants, such as gigantic cedars, pines, and redwood trees, that can live for many millennia. The amazing diversity of the eukaryotic photosynthesizers, almost all of which are likely descended from a single endosymbiotic event, is due to numerous changes in many aspects of plant and algal metabolism, physiology, and morphology since 2.3 Ga. In contrast, during the billions of years of its existence, the fundamental structure of the photosynthetic apparatus inside all extant plants and algae has changed little.

This leads to the important question: can we realistically expect to create improved versions of photosynthesis using modern biotechnology? As we saw in Chapter 2 (Bigger picture 2.1), in contrast to photosynthetic eukaryotes, the cyanobacteria, while keeping the same basic structure, have evolved many specialized variants of their photosynthetic apparatus in order to adapt to the many environmental and climatic changes they have faced over the past three billion years. This means that although the photosynthetic eukaryotes collectively outperform cyanobacteria in terms of their overall biomass and productivity, the cyanobacteria are arguable more versatile in terms of the range of ecological niches that they are able to inhabit. As we will see in Chapter 7, the newly discovered photosynthetic diversity of cyanobacteria might provide a much improved 'tool kit' to enable biotechnologists to modify eukaryotic photosynthesis. This could be one of the ways to address contemporary challenges such as: improving food security by increasing crop production in the face of climatic change; moving from fossil fuels to renewable bio-based energy production; and even the possibility of developing new technologies involving artificial photosynthetic systems.

The cyanobacteria comprise a large and successful phylum that constitutes one of the most diverse and widespread groups of bacteria dating back to the earliest stages of bacterial evolution. Recent phylogenetic evidence suggests that oxygenic photosynthesis might have been present in ancestral groups that led to cyanobacteria as early as 3.0 Ga (see Fig 4.2). There is some geological evidence showing that over the next 500 million years the activity of these oxygenic organisms led to localized 'whiffs' of oxygen in an otherwise anaerobic world as shown later in Fig 4.6. However, these results have been questioned and during most of the Archean the average oxygen levels were probably at least one million times less than those of today. By 2.5 Ga, it is likely that cyanobacteria were well established, most likely in low salinity and freshwater habitats such as brackish estuaries and deltas, and in terrestrial rivers and lakes. These oxygenic cyanobacteria could use the plentiful supplies of liquid water as a source of protons (H^+) and electrons (e^-) to drive their more productive form of photosynthesis. This also involved the emission of increasing quantities of gaseous oxygen as a by-product that had momentous consequences for the biology and geology of the Earth.

By about 2.4 to 2.2 Ga the amount of oxygen in the atmosphere suddenly (on a geological timescale) increased from near-zero to a global level of about 1% of the present value during the GOE1. As discussed in Chapter 1, the cause(s) of GOE1 and subsequent oxygenation events have yet to be conclusively determined. It has been widely assumed that such dramatic increases in net oxygen levels were due to the photosynthetic activity of cyanobacteria and the newly emerged photosynthetic eukaryotic organisms. However, recent modelling studies show that such oxygenations could also have been driven by global phosphorus, carbon, and oxygen cycles

Fig. 4.2 Estimates of the timing of the primary endosymbiosis event.

These estimates are based on molecular clock data from seven independent studies of the divergence times of the crown group of photosynthetic eukaryotes and the split of the plastid lineages from their cyanobacteria ancestors. The earlier studies tended to date the primary endosymbiosis event to between 1.6 and 0.9 Ga, while more recent studies using more comprehensive sequence datasets place the event at around 2.0 Ga, which is relatively soon after GOE1 resulted in a more aerobic world suitable for eukaryotic life. Circles and triangles are the mean ages for the origin of the crown group of photosynthetic eukaryotes and the common ancestor with cyanobacteria, respectively, and the rectangles represent the confidence intervals.

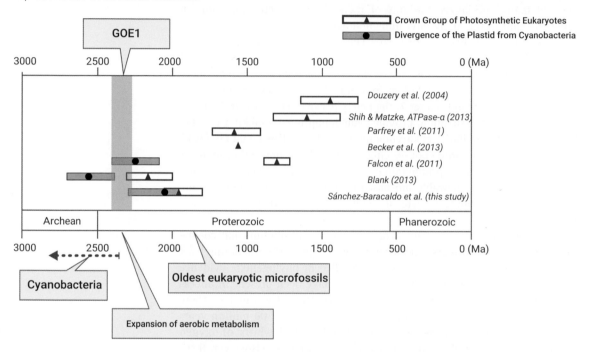

Image courtesy of Dr. Sánchez-Baracaldo, https://doi.org/10.1073/pnas.1620089114 Copyright © 2017

that only required the presence of simple photosynthetic cyanobacteria. Therefore the relative contributions of biology and geochemistry to GOE1 are still a topic of considerable research with new discoveries being made on a regular basis.

> The GOE1 transition created a completely new type of world that was increasingly dominated by a highly reactive gas, oxygen, that was toxic to the vast majority of extant life forms.

This was because the vast majority cellular life up to this time was almost exclusively made up of obligate anaerobes that were acutely sensitive to even small concentrations of oxygen and the resulting reactive oxygen species (ROS) that were generated in its presence. GOE1 laid the foundations for the rise of a hitherto obscure group of facultative aerobes. These were specialized, non-obligate anaerobes that had developed the capacity to tolerate the presence of small amounts of oxygen in their immediate environment. Even before GOE1, localized 'oases' of relative oxygen-rich habitats probably existed in the vicinity of regions where oxygenic photosynthesis by cyanobacteria were concentrated, even though most of the world would have remained essentially anoxic. As cyanobacteria proliferated and spread globally, the entire world was essentially 'engineered' away from a mainly anaerobic system (with a few aerobic 'oases') to one that was increasingly aerobic. Some anaerobic 'oases' remained and are still present today in places ranging from waterlogged soils to the human gut.

The organisms that benefited most from a more fully oxygenated Earth after GOE1 obviously included the cyanobacteria themselves. But it also included numerous groups of heterotrophic bacteria and, importantly, the recently evolved eukaryotes (see section 4.3), all of which were facultative aerobes able to withstand and even thrive within the increasing levels of oxygen in their environment. In contrast, obligate anaerobes became relegated to a dwindling number of habitats that remained anoxic. These residual anoxic regions included a restricted and disconnected range of subterranean, deep sea and volcanic habitats where a relatively small number of anaerobic organisms still thrive, although their biomass is far smaller that of the aerobic organisms that came to dominate most of the world. The stage was now set for the two biggest winners from GOE1, namely cyanobacteria and eukaryotes, to join together in a unique partnership, creating a biological 'killer app' that would go on to dominate the biosphere. This 'killer app' was formed when a cyanobacterium entered into an endosymbiotic association with a heterotrophic, aerobic eukaryote to produce the first eukaryotic organism with the capacity for oxygenic photosynthesis.

Comparisons of extant plastid- and nuclear-encoded genes of cyanobacterial origin in plants and algae with genomic data from cyanobacteria indicate that the original cyanobacterial partner in the endosymbiosis was most similar to the present-day freshwater species, *Gloeomargarita lithophora*. This is consistent with other evidence that the first photosynthetic eukaryotes might have evolved in a semi-terrestrial environment, such as in freshwater biofilms or microbial mats. Similar habitats are still home to many species of cyanobacteria and simple heterotrophic eukaryotes. Microbial mats form compact and relatively protected mini-ecosystems where metabolic symbioses and syntrophies between cyanobacteria and various heterotrophic eukaryotes can still be observed. According to the microbial mat scenario, the intimately close physical and genetic interactions between the two symbiotic, but physically separate, partners in the mat might have eventually led to the engulfment of the smaller partner, the cyanobacterial cell, by a much larger heterotrophic eukaryotic cell via the process of phagocytosis. In addition to the microbial mat scenario, there are several alternative hypotheses concerning the putative cyanobacterial ancestor(s) of plastids. These include the proposition that, instead of *G. lithophora*, a relative of a heterocystous cyanobacterium was the principal ancestor, or that alternatively several cyanobacterial species might have contributed to the endosymbiosis.

Normally the engulfment of a prey cell via phagocytosis would lead to its rapid digestion. However, as outlined below in section 4.4, we now know of several cases where such digestion does not happen immediately. For example, it has been shown experimentally that some eukaryotes are able to retain various types of engulfed bacteria inside their cytosol for extended periods. During this time, the bacteria can remain fully or partially functional, and they can also act as symbiotic partners to their host cell, albeit on a temporary basis until they are eventually digested. One can therefore speculate that, in some cases, if digestion of the engulfed cell were somehow inhibited, this could result in a more intimate endosymbiotic relationship between the two partners. This seems to have been a very rare event that only occurred a few times during the entire multi-billion-year course of biological evolution. The outcome was that one lucky cyanobacterial cell was retained and eventually 'domesticated' into a plastid that became a permanent part of the eukaryotic host cell.

Further evidence of the cyanobacterial origins of plastids comes from several structural features including their membranes. Cyanobacteria have a single cell membrane that is highly enriched in glycolipids, especially galactolipids. In contrast, eukaryotic cell membranes are typically enriched in phospholipids. As shown in Fig 4.3, when the cyanobacterium was engulfed via phagocytosis, it was further sealed into an additional second membrane derived from the plasma membrane of the host cell. This resulted in an endosymbiont that was surrounded by two membranes. The inner glycolipid membrane is derived from the cyanobacterium while the outer membrane has a mixed phospholipid/glycolipid composition that might be

Fig. 4.3 Outline mechanism of primary endosymbiosis.

In primary endosymbiosis, it is believed that an as-yet unknown type of heterotrophic eukaryotic cell engulfed a cyanobacterium that was similar to the present-day freshwater species, *Gloeomargarita lithophora*. Unusually, however, the cyanobacterium was not digested but was instead converted into a chloroplast organelle that was no longer capable of an independent existence. Similarly to mitochondria, which are also derived from an ancient endosymbiotic event, chloroplasts are surrounded by two membranes called the inner and outer envelope. See main text for more details.

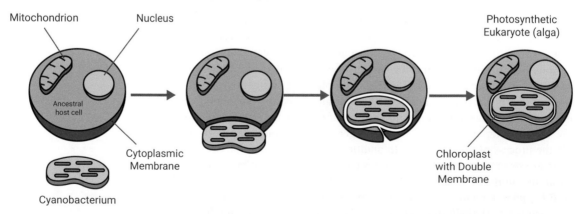

derived from both the host and the endosymbiont. Even today, primary plastids still retain the two distinctive membranes that resulted from their original engulfment by the host cell. As we will see in section 4.3, the cyanobacterial partner eventually lost its autonomy and was reduced to an existence as a mere organelle—the plastid—inside the much larger dominant host cell. As a consequence of this primary endosymbiosis, the eukaryotic host cell no longer needed to be an obligate heterotroph that had to 'hunt' for food in order to survive. Instead, it became an oxygenic autotroph that could manufacture its own food via sunlight and simple inorganic molecules, such as water, nitrates, minerals, and CO_2. Moreover, thanks to their aerobic nature, these new photosynthetic eukaryotes were also to flourish and diversify in an increasingly oxygenated world.

4.3   The eukaryotic partner

Unlike the relatively well characterized cyanobacterial ancestor of plastids described earlier in this chapter, we know remarkably little about the identity of the putative eukaryotic host cell. Evidence from the fossil record shows that recognizable eukaryotic cells were already present by 1.6 to 1.8 Ga. Early eukaryotes almost certainly existed well before this time but, because such cells would have lacked hard external coverings, it is unlikely that they were preserved as fossils. Phylogenetic studies of eukaryotic genomes imply that LECA could even date back as early as the start of the Proterozoic Eon, although an exact date for LECA remain highly controversial. As described in Chapter 1, there in increasing evidence that LECA was the result of the uptake of an α-proteobacterial cell by an archaeal cell similar to extant Asgard archaea. One possibility is that the ancestral eukaryote resulted from a symbiotic association between a hydrogen-dependent archaeon and a metabolically flexible α-proteobacterium. In the absence of oxygen, this adaptable bacterium would be able to survive and grow anaerobically, generating hydrogen as a side product. In the presence of oxygen, the bacterium could also function as an aerobe capable of oxidative respiration. Therefore, the ancestral eukaryote is likely to have been a facultative anaerobe, able to live in both oxic and anoxic environments.

There are several different theories about how the fusion of bacterial cell with an archaeal cell could have led to the formation of the first eukaryotic cell. We will consider two of these possibilities here (see Fig 4.4). The first theory is that, unlike the endosymbiotic event that produced the first algae, the bacterial/archaeal fusion was a much more even-handed process with no dominant host cell. Instead, most of the genomes of the two partner cells were eventually merged into what became the nucleus of a new type of composite cell and what was left of the bacterium became a new organelle, namely the ATP-generating mitochondrion. Mitochondria still contain bacterial-like 70S ribosomes, and have a permeable outer membrane complete with porins similar to those of contemporary aerobic bacteria. While it quickly lost most of its genes to the nucleus, the bacterial-derived mitochondrion retained a few dozen genes that were organized on a bacteria-like circular chromosome with a single origin of replication. In contrast, the rest of the genes in the newly formed nucleus had

Fig. 4.4 Two alternative models for the origin of the first eukaryotes (LECA).

The symbiosis model on the right proposes the direct fusion of an archaeal cell with a bacterial cell. How this endosymbiosis took place is not clear yet. Phagocytosis is the most likely mechanism for uptake of the endosymbiont, but there is no experimental evidence for this. The autogenous model on the left shows proto-eukaryotic cells arising from a common ancestor with archaea by progressive increases in complexity to became similar to modern eukaryotes. These cells had an endomembrane system, a nucleus, and a cytoskeleton, but no mitochondria. They were able to feed by phagocytosis and one of them engulfed an α-proteobacterium that was not digested. This ingested cell became an endosymbiont that was the ancestor of today's mitochondria. Many genes from the endosymbiont were transferred to the nucleus and took the control of most cellular metabolism, but not DNA processing which remained similar to the original archaeal host cell. As discussed in Chapter 1, recent evidence favours the symbiosis model, which also has the virtue of simplicity.

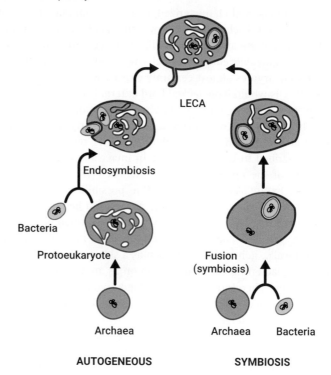

LECA

Endosymbiosis

Bacteria

Protoeukaryote

Fusion (symbiosis)

Archaea

Archaea   Bacteria

AUTOGENEOUS   **SYMBIOSIS**

multiple origins of replication, with DNA-binding histone proteins and a non-bacterial type of gene transcription. These are all archaea-like features that are not found in any bacteria, but are now core elements in the organization of all eukaryotic genomes. The second theory is that eukaryotes arose after two separate events. Firstly, an initial bacterial/archaeal fusion that produced a proto-eukaryotic cell that still lacked a mitochondrion. Secondly was the endosymbiotic capture of an α-proteobacterial cell by the proto-eukaryote that led to a fully-fledged eukaryote that contained a functioning mitochondrion.

Whatever the exact mechanism, there is robust evidence that eukaryotes arose following one or more fusion/partnerships between (i) an archaeal cell similar to present day Asgard group, the Lokiarchaeota, and (ii) a bacterial partner(s), one of which was similar to a present day α-proteobacterium and was the progenitor of mitochondria. Although much of their genetic machinery was of archaeal origin, the new eukaryotic organisms were far from being mere glorified archaea. Even though they now had an archaea-like organization, the majority of the nuclear genes were actually of bacterial origin. Also, all of the membranes of the new eukaryotic cells, including their novel endomembrane systems (e.g. endoplasmic reticulum (ER), peroxisomes, vacuoles, and nuclear envelope), were made up of bacterial-style phospholipid esters, instead of the phospholipid ethers that make up all archaeal membranes. The new eukaryotic organisms were genuine chimeras—ie they were a mixture of two radically different types of cell where significant attributes of each of the parental cells were retained. It is likely that all of the subsequent eukaryotic lineages, from humans to fungi, and from algae to oak trees, are derived from one small population of composite cells derived from archaeal and bacterial progenitors. This small group of cells then gave rise to LECA, which probably already contained mitochondrial organelles.

As oxygen levels continued to rise in the oceans, the chimeric eukaryotes with their mitochondrial 'power plants' gradually established themselves in a changing world. Eukaryotes proved to have several adaptive advantages over the prokaryotic bacteria and archaea that had been the only forms of life up to this time. In particular, eukaryotes were capable of highly efficient aerobic respiration, in contrast to prokaryotes, all of which lacked mitochondria. The earliest eukaryotes were probably basically anaerobic, but with some tolerance of oxidative stress, and they quickly developed improved mechanisms to cope with increasing levels of such stress as oxygen concentrations continued to rise. There is evidence that LECA already had some features that enabled it to survive in increasingly aerobic environments, including free-radical scavenging systems such as superoxide dismutase (SOD). This was especially advantageous because about 2.2 to 2.4 Ga, during GOE1, global oxygen levels rose by several orders of magnitude over a period of as little as one to ten million years as depicted in the model in Fig 4.5.

By about 2.0 Ga the Proterozoic world had started a momentous transformation from being a highly anaerobic place dominated by relatively small and slow-growing bacterial and archaeal prokaryotic cells into an increasingly productive aerobic environment that provided new evolutionary opportunities for two particular types of organism. These were the (i) oxygenic cyanobacteria that had initiated and were sustaining the ongoing global rise in oxygen levels and (ii) the (relatively) newly emerged eukaryotes. At this stage, the latter were all heterotrophs and, thanks to their mitochondrial organelles and more sophisticated cellular and genomic organizations, they were capable of more rapid growth and had much larger cell sizes than

Fig. 4.5 Evolution of Earth's atmospheric oxygen content through time.

For most of the Proterozoic Eon, free O_2 was possibly around 10%–20% of present atmospheric level (PAL), with deep ocean dissolved O_2 concentrations at about 11 mM or roughly 6% the present value of 178 mM. Towards the end of the Phanerozoic O_2 levels rose during the GOE2 until they reached, and even exceeded present levels. 'Whiffs' refer to isotope signatures for evidence of local O_2 before the GOE1. The Lomagundi excursion is a transient rise in O_2 occurring about 2.4 to 2.3 Ga. Heterocysts refers to differentiated cells produced by some cyanobacteria to protect nitrogenase from inactivation by O_2. Their relevance is that cyanobacteria have an ancient fossil record, but the oldest fossil heterocysts are younger than land plants, suggesting that cyanobacteria evolved this mechanism of O_2 protection in response to GOE2.

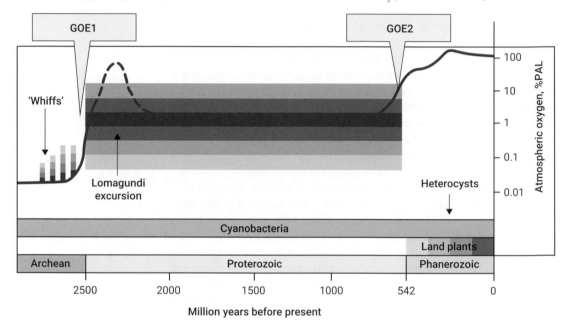

any of the prokaryotes. These early eukaryotes also developed ever more effective mechanisms for coping with oxidative stress, including additional enzymatic antioxidants such as catalase, ascorbate peroxidase and gluta-thione reductase. As we will now see, the uptake of a cyanobacterium by one of these eukaryotic cells created the highly diverse lineage called the Archaeplastida (see Fig 4.6). This group includes all of the algal and plant groups that make up about 85% of today's entire global biomass.

4.4   Primary endosymbiosis: mechanism and timing

Soon after GOE1, at about 2.0 Ga in the Proterozoic Eon, the increasingly oxygenated aquatic environment created an opportunity for the evolution of organisms that were able to adapt to more highly aerobic conditions. Such organisms included the relatively recently emerged archaeal/bacterial chimeras, the eukaryotes. As discussed in Chapter 1, an archaeon called

Fig. 4.6 The Archaeplastida are derived from primary endosymbiosis.

The primary endosymbiosis gave rise to a single cell that was the ancestor of the three major groups of the Archaeplastida algae. These groups are the glaucophytes, rhodophytes (red alage), and chlorophytes/streptophytes (green algae). The streptophytes include the ancestors of the land plants). The plastids of all of these organisms are surrounded by an inner membrane (im) and an outer envelope membrane (om). Nuclei shown as purple circles and mitochondria as purple ovals with hatched lines.

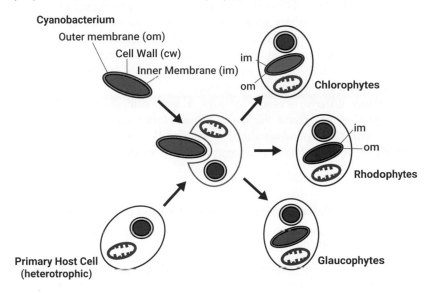

Prometheoarchaeum with long, branching protrusions able to ensnare other organisms was described in 2020. It is therefore suggested that an ancient relative of *Prometheoarchaeum* could have ensnared a bacterial cell to create the first eukaryotic cell. Eukaryotes soon developed comparatively large cells that were capable of taking up smaller prey cells, such as bacteria, by means of phagocytosis and digesting them in modified food vacuoles derived from the endoplasmic reticulum (ER). Therefore, heterotrophic eukaryotic cells were routinely using phagocytosis to consume a variety of cells that included cyanobacteria. In the case of the original or 'primary' endosymbiotic event that led to plastids, an interesting question is: why was the engulfed cyanobacterium was not promptly digested as would normally be the case for a prey cell?

   As shown in Fig 4.3, the engulfed cyanobacterial cell remained intact inside the eukaryote and continued to photosynthesize. There are clues to how this happened from studies of extant organisms that have ingested, but not digested, photosynthetic cells. For example, some cells of the heterotrophic eukaryote, *Paramecium bursaria*, contain endosymbiotic cells from *Chlorella*, a eukaryotic alga. This is not an obligate relationship as *P. bursaria* cells without algal endosymbionts are fully viable as heterotrophs.

Other eukaryotes are able to engulf and 'steal' individual plastids from algae, a process called kleptoplasty. An example of kleptoplasty is the non-photosynthetic ciliate, *Myrionecta rubra*, which regularly acquires plastids from cryptophyte algal prey. In kleptoplasty, the eukaryotic cells digest most of the algal prey cell contents, but retain their plastids intact for extended periods. However, in such cases, the ingested plastids eventually either degrade and/or are digested. Therefore, the 'host' cell needs to continually find additional algal prey to replenish its supply of plastids. Although these examples are not true long-term endosymbioses similar to the extremely rare event that led to plastid organelles, they nevertheless demonstrate that functioning photosynthetic cells and even individual organelles can be taken up and maintained intracellularly for prolonged periods by eukaryotes without being digested.

A rather different example of such a symbiotic process is found in the aquatic fern, *Azolla filiculoides*, which forms an extracellular symbiosis with the cyanobacterium, *Nostoc azollae* (see Fig 4.7). Because the *Azolla* fern is already photosynthetic, it does not rely on its resident

Fig. 4.7 Extracellular symbiosis between a fern and a cyanobacterium.

The host plant, the fern, *Azolla filiculoides,* houses its cyanobacterial symbionts, *Nostoc azollae,* in leaf cavities but these are outside the leaf cells. Because they are extracellular, these cyanobacteria are not true endosymbionts, although they have lost several key genes and are unable to grow outside the host plant. Unlike primary plastids of algae, the main function of these cyanobacterial symbionts is nitrogen fixation rather than photosynthesis.

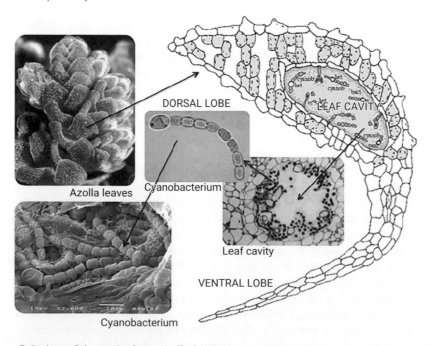

Azolla leaves · Cyanobacterium · DORSAL LOBE · LEAF CAVITY · Leaf cavity · VENTRAL LOBE · Cyanobacterium

cyanobacterial cells for photosynthesis, but it does benefit from their ability to fix atmospheric nitrogen. In this case, although the cyanobacterial symbiont remains outside the cells of its host, it has lost several critical genes involved in glycolysis and nutrient uptake and so is incapable of autonomous growth. However, the cyanobacterial cells remain intact and are passed on to subsequent generations of the plant host as internal (inside the host plant) but extracellular obligate symbionts. Although it has not developed into a fully-fledged endosymbiosis, this co-evolutionary association between a land plant and a cyanobacterium has been very successful and probably dates back to at least 140 Ma. Interestingly, recent analysis of the *Nostoc* genome suggests that it is still undergoing a process of gene loss and deactivation and that this might eventually cause it to resemble a plastid-like organelle rather than a bacterial cell. In addition to the fern, *Azolla*, *Nostoc* has formed similar nitrogen-fixing symbiotic partnerships with several other plant groups including liverworts, hornworts, cycads, and members of the angiosperm genus, *Gunnera*.

The above examples from *Paramecium* and *Myrionecta* show that some eukaryotes can ingest photosynthetic cells without digesting them and then go on to form temporary endosymbiotic associations with these cells. In the case of the original endosymbiotic event that led to primary plastids, the major difference was that the association was permanent. The two cells entered into a symbiotic association whereby the eukaryotic host cell provided a secure habitat and nutrients, including CO_2 and nitrates, for the cyanobacterium. In return, the cyanobacterium used light to generate complex molecules such as sugars, amino acids, and lipids for the host cell. Subsequently the cyanobacterium evolved into a metabolically much-reduced photosynthetic organelle known as a primary plastid. As part of this process, the vast majority of the original cyanobacterial genes were either completely lost or were transferred intact and functional to the host nuclear genome. It is estimated that up to 20% of the genes in extant algal and plant nuclear genomes are derived from their original cyanobacterial endosymbiont. During the transfer process these bacterial genes were modified in order to function in a eukaryotic genetic system. This involved adaptations such as different codon usage requirements, new regulatory sequences, acquiring protein localization signals, integrating host transcription and translation signals, and the incorporation of introns. An example of a contemporary unicellular green algal cell with its large endosymbiotic plastid organelle is shown in Fig 4.8.

Given the complexity of the process of achieving a successful primary endosymbiosis between a eukaryote and a cyanobacterium, it is not surprising that it is an extremely rare event. Indeed, until recently, it was thought that it had probably occurred only once, when the endosymbiotic event described above led to the origin of the Archaeplastida. However, a second example has recently been found in a genus of rhizarian euglenoids. This refers to *Paulinella chromatophora*, an alga that consists of eukaryotic amoeboid cells containing plastids descended from a cyanobacterial endosymbiont related to the contemporary free-living genus, *Synechococcus*. Using molecular clock methods, the original endosymbiotic event has been dated to between 140 and 90 Ma, or possibly somewhat earlier, which is considerably more recent than the event that led to the Archaeplastida. The

Fig. 4.8 Structure of a green algal cell from *Chlamydomonas reinhardtii*.

This primary algal cell is 5-10 μm in diameter with a single large chloroplast and two anterior cilia that are critical for movement and mating. These motile algae also have a pigmented eyespot that allows them to sense light. The contractile vacuoles are used to expel surplus fluid from the cell. The pyrenoid is a compartment found in many algae and in a single group of land plants, the hornworts. Its function is to increase the local concentration of CO_2 near the active site of rubisco, as discussed in Chapter 3.

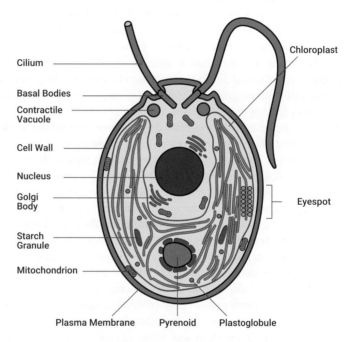

Source: S. Sasso, H. Stibor, M. Mittag, A. R. Grossman, 'The Natural History of Model Organisms: From molecular manipulation of domesticated Chlamydomonas reinhardtii to survival in nature' in eLife Sciences, 7:e39233, 2018, fig. 1. Licensed under CC BY 4.0. DOI: 10.7554/eLife.39233 [link source URL to DOI]

relatively recent timing of the *Paulinella* endosymbiosis is consistent with estimates that its plastids have only lost about two-thirds of their original genes, whereas in the Archaeplastida the figure is more like over 95% of plastid genes having been lost.

In the case of the main primary endosymbiosis that led to the Archaeplastida, both genomic and fossil evidence date the event at 2.0 to 1.8 Ga and possibly earlier.

For example, there are well preserved fossils of the red alga, *Bangiomorpha*, from 1.0 Ga and it is likely that green and red algae diverged several hundred million years beforehand, with potential green algal fossils dating from as early as 1.8 Ga. Other algal-like fossils, named as *Rafatazmia* and *Ramathallus,* have been interpreted as being red algae dating from 1.6 Ga. Well-preserved cells of a giant algal-like unicellular organism called *Grypania* have been dated from about 1.8 to 2.1 Ga. If, as is claimed by some researchers, these fossils are indeed from green algae, this would suggest that the original primary endosymbiosis could have occurred as early as 2.3 to 2.2 Ga, which is close to the timing of GOE1. However, it should be mentioned that there is an alternative interpretation that *Grypania* might be a giant cyanobacterium rather than an alga, so these timing estimates cannot be regarded as definitive at present.

To summarize, the fossil evidence for evolution of the early algae remains incomplete, and in some cases is still controversial, but it is in general agreement with independent lines of evidence from genomic studies. Together, both lines of evidence suggest that the primary endosymbiosis that led to the Archaeplastida probably occurred relatively soon after GOE1 at around 2.0 Ga +/- 0.3 Ga. The appearance of putative red and green algal fossils by 1.6 to 1.8 Ga suggests that the cyanobacterial endosymbiont was converted into a plastid organelle with a reduced genome relatively soon after the primary endosymbiosis event, possibly within 100 to 200 million years. As discussed in more detail in Chapter 5, the primary endosymbiosis was followed by an enormous diversification of the major algal groups within the Archaeplastida during the mid-Proterozoic Eon.

4.5   Plastid evolution and diversification

The primary endosymbiosis event involved the uptake of a fully intact and functional cyanobacterial cell that subsequently became drastically modified to produce the plastid organelles seen in all algae and plants.

The word 'plastid' is derived from the same Greek root as 'plastic', with the meaning of having a flexible shape. As we will now see, in contrast to their cyanobacterial ancestor, plastid organelles have evolved into dozens of different shapes and sizes, while still carrying the remains of their bacterial heritage within their much-reduced genomes. Whereas the genomes of free-living cyanobacteria contain about 4,000 to 6,000 protein-encoding genes (about 5 to 8 Mb of DNA), their contemporary plastid descendants contain only 160 to 235 such genes in the case of extant red algae, and only 100 to 220 genes in extant green algae and plants (about 100 to 200 Kb of DNA). This massive gene loss was an important part of the mechanism whereby the original endosymbiont was irreversibly converted into a dependent organelle.

The loss of a few key genes early in this process could have 'fixed' the cyanobacterial cell into dependence on its eukaryotic host. For example, glycogen synthase genes, which are not found in plastids, might have been some the first genes to be lost from the cyanobacterial endosymbiont. In the absence of any ability to store photosynthate as glycogen, intermediates such as simple sugars would have leaked out of the cyanobacterial cell, dramatically reducing its viability as an independent entity while greatly benefiting the host cell. The loss of the ability to synthesize glycogen was accompanied by a switch to starch as the major storage carbohydrate in the ancestor of the Archaeplastida. In glaucophytes and red algae, starch is now stored in the cytosol and this was probably the original location of the process. However, soon after their divergence from glaucophytes and red algae, the green algae relocated starch biosynthesis to the plastid, where it is still found today in their plant descendants.

While many of the genes in the original cyanobacterial genome were lost, a significant number were transferred in an intact and fully functional state to the host nucleus. Most of these transferred genes are involved in photosynthesis-related functions. Examples include genes encoding ancillary photosystem subunits and chlorophyll biosynthetic enzymes, as well as aspects of plastid metabolism and maintenance, such as amino acid synthesis and plastid division. Recent proteomic studies have also identified an important group of transferred genes that are involved in redox sensing. The presence of these genes in the host cell led to the development of more effective redox-sensing pathways. These enabled the host cell to cope with increased levels of oxidative stress due to the localized release of ROS that were being produced by their newly acquired plastid organelles. The same studies have also identified several proteins in present-day plastids that are encoded by nuclear genes originating from bacteria that are different from the original cyanobacterial endosymbiont. This indicates the possible occurrence of Horizontal Gene Transfer (HGT) whereby additional genes were acquired at some stage during algal evolution.

The latter finding suggests that, even after acquiring its photosynthetic endosymbiont, the eukaryotic host cell still continued its heterotrophic lifestyle, at least for a while. The newly photosynthetic eukaryotes would have remained as facultative heterotrophs, ingesting bacterial prey cells and acquiring genes from them via HGT. Further evidence for this comes from the extant green alga, *Cymbomonas*, which is fully photosynthetic but has retained an ability to use phagocytosis to ingest bacterial prey in order to supplement its diet. Many Chlorophyte green algae have retained an ability to move around in their environment (motility), and many species, such as *Cymbomonas*, have a facultatively heterotrophic (or mixotrophic) lifestyle. In contrast, motility has been largely lost in most glaucophytes and red algae, as well as in the Streptophytes, which include the Charophyte green algae and the land plants. Therefore, all of the latter non-motile groups, including the land plants, are now obligate autotrophs. An exception to this is the evolution of a few groups of land plants that, as discussed in Chapter 6, have secondarily lost their ability to photosynthesize and have instead reverted to heterotrophic, mainly parasitic, lifestyles.

Following their 'domestication' by the host cell, plastids diversified into a range of different morphological structures in the various algal lineages as discussed further in Chapter 5. In the majority of cases, the plastids in extant algal cells are in the form of chloroplasts, ie they are green organelles that contain thylakoid membranes that house the photosynthetic pigment-protein complexes. In chloroplasts from red algae and glaucophytes, the thylakoid membranes additionally contain phycobilisomes, which are large pigment-protein arrays used for light harvesting as also found in cyanobacteria. In contrast, the thylakoids of green algal chloroplasts are devoid of phycobilisomes. As some algae increased in complexity and eventually became comparatively large multicellular organisms, their plastids became specialized to fulfil new roles in addition to photosynthesis.

This phenomenon is most marked in the green algae and plants. For example, in chlorophytes such as *Ulva* spp. (sea lettuces), some of the plastids have lost their thylakoid membranes and have instead acquired new metabolic functions, such as becoming specialized starch-storing organelles. These plastids are the colourless amyloplasts that typically contain numerous starch granules. In addition to being present in some algae, amyloplasts are especially common in storage tissues of land plants, including seeds, tubers, and rhizomes. In other multicellular chlorophyte algae, the cells in their rapidly dividing apical regions contain proplastids, which are small nascent plastids that have yet to differentiate into mature forms such as amyloplasts or chloroplasts. Proplastids are also found in the developing reproductive structures of some red algae, such as *Janczewskia gardneri*. Yet another type of plastid found in some of the more complex algae are chromoplasts. These are highly coloured, but non-green, organelles that accumulate lipidic globules made up of pigments such as β-carotenoids (orange), xanthophylls (yellow), and lycopenes (red). It should be stressed that all of these diverse plastid forms still contain their cyanobacterial-derived genome and continue to carry out several core metabolic functions such as fatty acid and starch biosynthesis.

At about 400 Ma, the evolution and diversification of plastid organelles was taken even further when one particular group of the Streptophyte green algae started to colonize the land. This monophyletic group was the Embryophytes, or land plants, and its members developed completely new types of organs and tissues, some of which are completely non-photosynthetic. Notable examples include subterranean organs such as roots, rhizomes, corms, bulbs, and stolons. Some Embryophytes also developed highly coloured organs, such as flower petals and fruits, as part their co-evolutionary relationship with various metazoan groups such as insect pollinators or mammalian fruit eaters. It is estimated that over half of the biomass of most land plants is made up of non-photosynthetic tissues. However, because all of the cells in these non-green tissues still contain plastids, these organelles must have acquired additional non-photosynthetic functions.

> In land plants, a young cell typically contains between 10 and 20 plastids, but this number can increase to over 100 in the case of highly active cells, such as mesophyll and bundle sheath tissues in the leaves of some higher plants.

During the life cycle of a plant cell, plastids are able to divide within the host cell so that approximately equal numbers can be passed on to the daughter cells after each cell division. However, the plastids in the many non-photosynthetic tissues of land plants that are not exposed to the light have no possibility to function as photosynthetic organelles. Hence they have developed numerous alternative functions, some of which, such as starch or pigment storage, were already present in their algal ancestors as discussed above. However, several new functions were also acquired, depending on the needs of the particular tissue in which the plastids are located (see Fig 4.9).

Additional examples of non-photosynthetic plastids found in land plants include the etioplasts, which are membrane-rich precursors of chloroplasts found in tissues that are normally green but have not been exposed to light; and the lipid-rich elaioplasts found in several colourless tissues. In maturing vanilla pods, chloroplasts lose their internal membranes and are converted to phenyloplasts. The stromal compartment of these colourless plastids becomes filled with the toxic phenylpropanoid, vanillin, which acts as a deterrent against herbivores. The chloroplasts in senescing leaves undergo a regulated process whereby their thylakoid membranes are broken down, with the glycolipids metabolized to triacylglycerols that accumulate in droplets termed plastoglobules. These former chloroplasts are termed gerontoplasts and the bulk of their lipid and protein contents is scavenged subsequently by the parent plant prior to leaf fall, or dehiscence. This process is especially marked in perennial plants that lose their leaves during

Fig. 4.9 Diversity and differentiation of plastids in plants and algae.

In many plants and some algae, plastids are able to differentiate into many forms, depending on the tissue type and developmental stage of the cells in which they are present. Arrows indicate possible transitions between different plastid types. Proplastids are the progenitor plastids, developing into either etioplasts or chloroplasts depending on exposure to darkness or light, respectively. Proplastids are also able to develop into the intermediate plastid form, leucoplasts, that may then further differentiate into amyloplasts, proteinoplasts, or elaioplasts accumulating starch, proteins, or oils, respectively. Chloroplasts can also further differentiate depending on environmental stimuli and plant cell types into chromoplasts, phenyloplasts and the senescing gerontoplasts, in which resources are recycled and redistributed to other cells. Chromoplasts and phenyloplasts accumulate phytochemicals such as carotenoids and phenylpropanoids, respectively. Some plant plastids also act as accumulation sites for antiherbivore defence compounds such as vanillin.

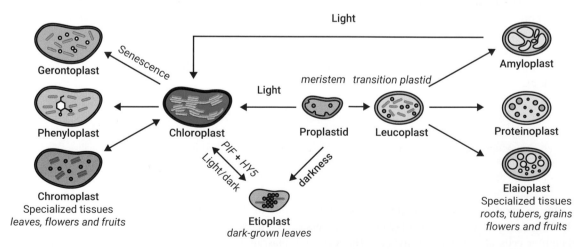

Dynamic metabolic solutions to the sessile life style of plants C. Knudsen, N. J. Gallage, C. C. Hansen, B. L. Møller and T. Laursen, Nat. Prod. Rep., 2018, 35, 1140 DOI: 10.1039/C8NP00037A

the autumn season in temperate regions. However, it also occurs through-out the year in non-deciduous plants as older leaves undergo senescence to make way for new leaves. A similar form of senescence can also occur following various forms of stress such as drought or ozone exposure.

None of the non-green plastids described above are capable of photosyn-thesis, but they are still bounded by exactly the same kind of double envelope membrane and carry an identical cyanobacterial-like genome to that found in chloroplasts. In addition to their photosynthetic and storage roles, all plas-tids perform several vital functions in plant cells. For example, in all plant cells the plastids are the only place where *de novo* fatty acid biosynthesis and various aspects of hormone and nitrogen metabolism can occur. Proplastids in plants have the potential to develop into any of the other forms of plastid, and in some cases, plastids can reversibly change from one form to another. For example, if developing leaves are moved from a light to a dark location their chloroplasts change into etioplasts, and if the same darkened leaves are put back in the light the etioplasts change back into chloroplasts. Like other aspects of their life cycles, such as division, plastid morphology is strictly under the control of genes located in the host cell nucleus.

In this chapter, we have seen evidence that a very rare and probably unique event led to the combination of a cyanobacterial cell with a much larger heterotrophic eukaryotic cell to form the first eukaryotes capable of oxygenic photosynthesis. This event is known as primary endosymbiosis and was the beginning of the evolution of all of the photosynthetic algae and plants, both terrestrial and aquatic. In the next chapter, we will look at how, during the one-billion-year period after GOE1, these diverse forms of photosynthetic eukaryotes gradually became more complex and radiated into a wide range of forms that occupied aquatic niches around the world. This set the scene for some of them to eventually move from their ancestral ecosystems to colonize a new and challenging habitat, namely dry land.

Chapter summary

- The original primary endosymbiotic event probably occurred between a photosynthetic cyanobacterium by a heterotrophic eukaryotic cell and resulted in the first photosynthetic eukaryote. The timing of this momentous event is uncertain but was plausibly about 2.0 Ga.

- The cyanobacterial endosymbiont was reduced from a free-living organism to an organelle, the plastid, that was completely under control of the host cell. As part of this process the vast majority of the plastid genome and most of its metabolic functions were transferred to the host nucleus.

- The new organism was now a primary algal cell, an oxygenic photosynthetic eukaryote. These highly efficient photosynthetic aerobic eukaryotes exploited more marine habitats, and benefited from increased oxygen levels and nutrient availability.

- By about 1.0 Ga algal rates of primary production and oxygen evolution overtook those of the cyanobacteria with the primary algae, or Archaeplastida, diversifying into several groups including the one that led to the land plants.
- Plastids continued to evolve in algae and plants, sometimes acquiring new non-photosynthetic functions related to storage, reproduction, or defence.

Further reading

Burki F, Roger AJ, Brown MW, Simpson AGB (2020) The new tree of eukaryotes, *Trends Ecol Evol* 35, 43–55. DOI: https://doi.org/10.1016/j.tree.2019.08.008
A current view of the eukaryotic family tree based on genomic and phylogenetic studies.

Gavelis GS, Gile GH (2018) How did cyanobacteria first embark on the path to becoming plastids?: Lessons from protist symbioses, *FEMS Microbiol Lett* 365, fny209. DOI: 10.1093/femsle/fny209
Tracing the journey of cyanobacteria from free-living organisms to plastid organelles.

Gibson TM et al (2017) Precise age of *Bangiomorpha pubescens* dates the origin of eukaryotic photosynthesis, *Geology* 46, 135–138. DOI: 10.1016/j.cub.2016.11.056
Geological evidence that a fossil multicellular red alga dates back to before 1.0 Ga.

Judson OP (2017) The energy expansions of evolution, *Nature Ecol Evol* 1, 138. DOI: 10.1038/s41559-017-0138
Outlines how energy from processes including sunlight, geochemistry, and oxygen have powered evolution on Earth.

Hehenberger E, Gast RJ, Keeling P (2019) A kleptoplastidic dinoflagellate and the tipping point between transient and fully integrated plastid endosymbiosis, *Proc of Nat Acad Sci USA* 116, 17934–117942. DOI: 10.1073/pnas.1910121116
Describes how a dinoflagellate 'steals' plastids from cyanobacteria, illustrating a possible route to endosymbiosis.

Mukherjee I, Large RR, Corkrey R, Danyushevsky L (2018) The boring billion, a slingshot for complex life on earth, *Scientific Reports* 8, 4432. DOI:10.1038/s41598-018-22695-x
Outlines how the period from 1.8 to 0.8 Ga was crucial for the evolution of complex life.

Ponce-Toledo RI, Deschamps P, López-García P, Zivanovic Y, Benzerara K, Moreira D (2017) An early-branching freshwater cyanobacterium at the origin of plastids, *Curr Biol* 27, 386–391. DOI: http://10.1130/G39829.1
Evidence that plastids originated from a cyanobacterium similar to extant Gloeomargarita.

Ponce-Toledo RI, Lopez-Garcia P, Moreira D (2019) Horizontal and endosymbiotic gene transfer in early plastid evolution, *New Phytol* 224, 618–624. DOI:10.1111/nph.15965
How modern plastids have inherited genes both from cyanobacteria and other bacteria.

Taverne YJ, Caron A, Diamond C, Fournier G, Lyons TW (2020) Oxidative stress and the early coevolution of life and biospheric oxygen, in: *Oxidative Stress: Eustress and Distress*, Sies H, ed, Academic Press, New York. DOI: 10.1016/B978-0-12-818606-0.00005-5
Outlines the crucial role of oxygen in the evolution of early life.

 ## Discussion questions

4.1 Describe the process by which a eukaryotic cell first acquired the ability to photosynthesize.

4.2 How was a free-living cyanobacterial cell transformed into a plastid organelle?

4.3 Describe how plastids have evolved within their host cells to acquire new functions apart from photosynthesis.

5 EVOLUTION OF THE ALGAE

Learning objectives

- Understanding how the algae ultimately originated from a single endosymbiotic event that led to the Archaeplastida group.

- Examining the diversification of the Archaeplastida into the three primary algal groups, namely red, glaucophyte, and green algae.

- Focusing on the further evolution of the Viridiplantae group that includes the green algae and their land plant descendants.

- Describing how numerous new algal groups arose due to **secondary endosymbiotic** events between red and/or green algae and a wide variety of heterotrophic protists.

- Learning how some algae lost the ability to photosynthesize and reverted to heterotrophy, in several cases becoming major human and crop pathogens.

5.1   Introduction

In Chapter 4, we saw how the engulfment of a cyanobacterial cell by a eukaryotic host cell led to the emergence of the first group of photosynthetic eukaryotes, the primary algae. During the subsequent process of endosymbiosis, the original cyanobacterial cell lost its autonomy and became reduced to the status of a dependant plastid organelle inside its host cell. For these newly emerged photosynthetic eukaryotes, adaptations to high levels of oxygen stress were required at two levels. The first level was within the broader external environment, where atmospheric and aquatic oxygen levels had risen by several orders of magnitude during the 1st Great Oxygenation Event (GOE1) that probably slightly predated the appearance of algae. The second level was inside the algal cells, where their newly acquired plastids were generating massive local concentrations of ROS during peak periods of photosynthesis. These two linked factors created unprecedented levels of oxidative stress that required rapid quenching of the ROS. This in turn required the deployment of much more effective anti-oxidant systems than those had existed previously. The emergence of such antioxidant mechanisms was a key factor that ensured the survival and further evolution of photosynthetic eukaryotes.

> The original photosynthetic eukaryotes were all relatively simple unicellular, aerobic organisms.

However, they quickly diversified into numerous species of both marine and freshwater algae, with some groups gradually becoming truly multicellular. The descendants of the original primary endosymbiosis event are known as the Archaeplastida. This large group of organisms includes all the primary red and green algal species that still make up much of the phytoplankton that sustains the major marine food webs from whelks to whales. The largest group in the Archaeplastida is the Viridiplantae, which includes the green algae from which multicellular land plants arose. The land plants are direct descendants of one group within the Viridiplantae called the Streptophytes, which began to colonize the Earth's surface after 1.0 Ga. As discussed in Chapter 6, the land plants include many large and complex photosynthetic organisms that ultimately sustain contemporary terrestrial food webs.

As the various groups of primary algae diversified, the proliferation and increasingly widespread distribution of these relatively fast-growing eukaryotic photosynthetic autotrophs provided a rich food source for other opportunistic organisms. Eventually, this led to the evolution of other complex multicellular heterotrophic organisms, such as animals and fungi. However, as shown in Case study 5.1, the initial endosymbiotic event that created the first photosynthetic eukaryotes did not immediately result in these dramatic changes. Indeed, it took over a billion years from the origins of the first relatively simple aquatic photosynthetic eukaryotes at about 2.0 Ga until they gradually became more complex and radiated across the world by about 1.0 Ga. From this time, however, they went on to develop into the huge diversity of algal and plant species that underpins and dominates the ecology of the contemporary biosphere.

5.2   Diversification of the Archaeplastida

The primary endosymbiosis event that gave rise to the green algae probably occurred in the early Proterozoic Eon, about 2.0 Ga. As discussed in Chapter 4, it is likely that the primordial algal cell was able to convert its newly acquired cyanobacterial endosymbiont into a plastid organelle relatively quickly. This ancestral alga then diversified into the three main lineages that make up the Archaeplastida, namely red algae, green algae, and glaucophytes. Fossils dating from 1.6 Ga in the Gaoyuzuang Formation in Northern China have been interpreted as the remains of very early Archaeplastida groups, that emerged prior to the divergence of the red and green algal lineages. Recognizable fossils similar to modern green and red algae have been dated to 1.6 to 1.8 Ga. At present, the precise timing

Case study 5.1
The significance of algal photosynthesis

The algae are of enormous significance in the evolution of photosynthesis and indeed of life in general. By about 2.3 Ga, the massive release of oxygen by cyanobacteria resulted in the GOE1. This altered the balance of life from being mostly anaerobic to being overwhelmingly aerobic. These aerobic conditions favoured the rise of a new group of organisms, the eukaryotes, that were probably the result of a fusion between an archaeal cell and an aerobic bacterium. The bacterium was converted into a mitochondrial organelle capable of highly efficient oxidative respiration that enabled its host cell to grow much faster and larger than either its archaeal or bacterial progenitors. It was one of these dynamic heterotrophic eukaryotic cells that engulfed a cyanobacterial cell and converted it into a chloroplast, hence giving rise to the first algal cell.

All extant non-bacterial photosynthetic organisms including all of the algae and plants, (with one exception—see Chapter 4), are probably descended from a single endosymbiotic event that occurred sometime after GOE1. Over the next billion years, the primary algae gradually increased in numbers and diversity to colonize marine and freshwater habits around the world. Due to their more efficient photosynthesis, rapid growth rates, and increasing complexity the algae eventually overtook cyanobacteria as the major global primary producers. By 1.0 Ga, some algae had diversified into the first truly multicellular organisms including the precursors of extant seaweeds. Other algal types evolved independently, following a series of secondary endosymbioses that gave rise to several ecologically important and highly photosynthetically productive groups, such as diatoms and dinoflagellates.

By about 0.8 to 0.6 Ga, the ever-increasing photosynthetic productivity of the algae led directly to the GOE2. This established levels of atmospheric oxygen comparable to those of the present day. The next great contribution of the algae was therefore to enable the evolution of complex multicellular organisms that were able to colonize the land. These included plants, animals, and fungi. Their success was made possible by the higher levels of atmospheric oxygen after GOE2, and establishment of a more effective ozone layer to screen out harmful wavelengths of solar radiation. Some groups of multicellular green algae were among the earliest terrestrial colonizers, where they successfully adapted to the new environment that included challenges such as more intense sunlight and reduced availability of water. As discussed in Chapter 6, it was from these Streptophyte algae that all extant land plants arose. In conclusion, the algae, with their 'turbo-charged' capacity for oxidative photosynthesis, were ultimately responsible for the evolution of all aquatic and terrestrial complex life on Earth.

and order of the appearance of each of the three Archaeplastida lineages is unclear. Some studies show red algae appearing first while others show that glaucophytes were first. However, it is probable that the Archaeplastida had already diverged into its three daughter groups by about 1.5 Ga, and that by 1.0 Ga more complex multicellular forms were emerging.

Glaucophytes and red algae

Interestingly, the plastids of glaucophytes and red algae have retained a greater number of cyanobacteria-like characteristics than the green algae. For example, both glaucophyte and red algal plastids contain unstacked thylakoid membranes with large protruding light-harvesting phycobilisomes arrays, similar to those found in cyanobacteria. Glaucophytes and red algae also lack chlorophyll *b* and synthesize their starch in the cytosol. In contrast, green algae have stacked thylakoids that lack phycobilisomes, although they do not contain granal membranes, which are unique to land plants. Green algae also contain chlorophyll *b*, and their starch biosynthesis occurs within their plastids rather than in the cytosol. These structural and metabolic differences support an earlier divergence of glaucophytes and red algae and their retention of some of the original cyanobacterial traits. Meanwhile, the green algae have more similarities to modern land plants than to the other two groups of Archaeplastida. These differences are also mirrored in their respective genome sequences. For these reasons, it is now recognized that the green algae and land plants together form a separate group, known as the Viridiplantae, that is quite distinct from the two other groups of Archaeplastida algae.

The glaucophytes are the smallest, and possibly the earliest diverging, of the three groups within the Archaeplastida (although this has been disputed). A variety of genomic studies places the most likely origin of the glaucophytes as early as 1.5 Ga. Given the small size and relative fragility of glaucophyte cells, it is likely that these algae were already present during the mid-Proterozoic, from 1.0 to 1.6 Ga, but were not preserved in the fossil record. Glaucophyte plastids are unique among the algae in being surrounded by a cyanobacteria-like vestigial peptidoglycan wall. Today, glaucophyte algae are very rare and are mainly of scientific interest as part of studies into plastid origins and evolution. Unlike other primary algae, they are restricted to freshwater habitats and are only found as simple unicellular or colonial coccoids or flagellates.

The red algae are a much larger and more evolutionarily successful group than the glaucophytes. They include several large and well-known multicellular species, such as Irish moss (*Chondrus crispus*, see Fig 5.1), dulse (*Palmaria palmata*), and laver (*Porphyra* spp). Today, red algae occupy many diverse marine, coastal, and freshwater habitats across the world. Fossil evidence from 1.6 Ga suggests that red algae were some of the earliest eukaryotes to become multicellular. One advantage that red algae have, compared to green algae, is the ability of their phycocyanin and phycoerythrin pigments to absorb a wider portion of the visible light spectrum. As shown in Fig 5.2, phycoerythrin and phycocyanin maximally absorb light at about 450–570 nm and 570–620 nm respectively. In contrast, the

Fig. 5.1 Examples of multicellular red algae.

Unlike many of the more advanced green algae, multicellular red algae remained relatively small and never colonised the land. **(a)–(d)** various forms of *Chondrus crispus*. **(e), (f)** various forms of *Gigartina mamillosa*. **(g)** Longitudinal section through fruit capsule of *G. mamillosa*. **(h)** Cross-section through a fruiting lobe of the thallus of *C. crispus*. **(i)** cross-section of fruiting *C. crispus* cystocarp. **(j)** the same, but much enlarged.

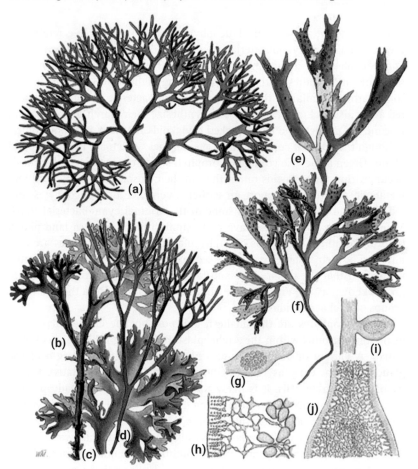

Source: Franz Eugen Köhler, 1887 / Wikimedia Commons / Public Domain

major green algal pigments, chlorophylls *a* and *b* and carotenoids, absorb maximally between 400–480 nm with a smaller peak at about 650–680 nm. The ability of red algae to harvest light in the mid-range of the visible spectrum is especially important at lower water depths where the shorter light wavelengths preferentially used by green algae are highly attenuated, while those used by red algae are still readily available. This means that, while green algae can only photosynthesize relatively close to the water surface, red algae can be found down to depths of as much as 200 metres.

Fig. 5.2 Absorption spectra of major algal and plant pigments.

Most land plants rely primarily on chlorophylls *a* and *b* and carotenoids but these pigments absorb very little light in the middle of the visible spectrum, from 500 to 640 nm. Some algae contain two additional pigments, phycoerythrin and phycocyanin, that enable them to harvest light in this middle part of the spectrum. This is especially useful in deeper regions of aquatic environments where shorter light wavelengths are highly attenuated.

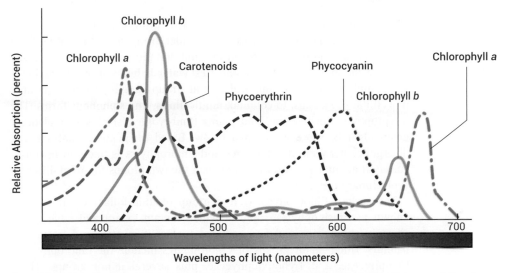

from *Microalgae: The Multifaceted Biomass of the 21st Century* By Donald Tyoker Kukwa and Maggie Chetty Published: December 16th 2020 DOI: 10.5772/intechopen.94090

Despite their development of multicellularity and the ability to occupy deeper aquatic niches, none of the red algae became particularly large or structurally complex. Neither were they able to move into terrestrial habitats. This sets the red algae apart from other major multicellular lineages, such as metazoans (animals), fungi, and Viridiplantae. There are some clues from genomics about the reason this evolutionary constraint in red algae, namely the absence of several cytoskeleton-related gene families. In particular, red algae lack dynein motor proteins and most also lack myosin. The lack of myosin is particularly striking because this protein is near-ubiquitous in eukaryotes and was almost certainly present in LECA. The data suggest that cytoskeleton-related genes were lost from the genome of an early ancestor of all extant red algae. The resultant structural deficiencies might explain the limited ability of red algae to form large cells and large, robust multicellular bodies when compared with many other eukaryotic groups. For example, the largest cells in the red alga, *Porphyra*, are thread-like, slow-growing rhizoid cells that are only a few millimetres long. In contrast, much larger cells, between 10 and 100 mm in length, with numerous specialized functions, have evolved in many freshwater and marine green algae and land plants, as well as in both terrestrial and aquatic animals and fungi.

The Viridiplantae

In terms of their diversity, global distribution, and overall biomass, the Viridiplantae are far more successful than either the glaucophytes or the red algae.

The Viridiplantae are divided into two groups, Chlorophytes and Streptophytes. The Chlorophytes are the older group, probably dating from about 1.4 Ga, and include most of the traditionally recognized green algae that are found in most marine and freshwater habitats. The Chlorophytes are a highly successful and widespread group ranging from planktonic unicellular organisms, to colonial, multicellular, and siphonous forms. The Streptophytes include the land plants (Embryophytes) plus several other green algal lineages as outlined below. Based on genome analyses, the divergence between Streptophytes and Chlorophytes is likely to have occurred at about 1.2 to 1.0 Ga after which the two groups followed different evolutionary trajectories.

The earliest Chlorophytes were a group of unicellular oceanic species known as Prasinophytes, some of which remain extant and have diversified in the oceans and also radiated further in coastal and freshwater habitats. Other major groups of Chlorophytes include the Chlorophyceae, Ulvophyceae, and Trebouxiophyceae, plus several minor groups. These latter groups radiated after the Prasinophytes, possibly after 1.0 Ga. The Ulvophyceae include mostly macroscopic, multicellular, or siphonous species in coastal habitats, plus many microscopic unicellular or larger multicellular species found in marine, freshwater, and terrestrial environments. One particularly prominent member of the Ulvophyceae is the multicellular species, *Ulva lactuca*, also known as the sea lettuce. This relatively large, edible seaweed is a common feature of coastal regions of Northern Europe where it flourishes particularly in intertidal ponds.

The reasons for the radiation of the major groups of green algae in the latter part of the Proterozoic remain unclear. However, recent data from the analysis of trace elements in Proterozoic oceans suggest that there was a significant increase in nutrient concentrations after 1.4 Ga. This is linked to a diversification of marine life, and especially of green algae, from about 1.4 to 0.8 Ga. During the early part of this timeline all of the main groups of green algae had already diverged and were gradually expanding into new oceanic habitats. It now seems likely that the free-living and syntrophic cyanobacteria were the major contributors to photosynthesis and, especially to oxygen evolution, during much of the Proterozoic. It is possible that nutrient deficiency, especially phosphate limitation, affected green algal growth capacity during this time and could have been an important contributor to cyanobacterial domination during much of the Proterozoic. With their smaller size and higher surface-to-volume ratios, cyanobacteria were more efficient at diffusion-controlled uptake of relatively scarce nutrients compared to the larger-celled algae, which would have been more developmentally constrained by such limitations.

This cyanobacterial dominance of global primary production gradually decreased as the availability of trace nutrient increased due to geological changes. As a consequence, all the main algal groups, but especially green algae, began to diverge and occupy new habitats during the latter part of the Proterozoic. As discussed above, red algae were also diverging at this time and, thanks to their phycocyanin and phycoerythrin pigments, they were able to occupy deeper marine niches than most green algae and cyanobacteria. Therefore, at about 1.2 to 1.0 Ga, several of the major groups of oxygenic primary algae were steadily increasing in abundance while that of the cyanobacteria was probably decreasing. Proxy evidence for such a decline comes from one of the major cyanobacterial habitats, namely the layered rock formations called stromatolites. From about 1.2 Ga the abundance of species in stromatolites went into a steady decline that is indicative of an equivalent reduction in overall cyanobacterial biomass (see Fig 5.3).

There is fossil evidence of a multicellular green alga, named *Proterocladus*, dating from 1.0 Ga. This suggests that green algae were already diversifying well before the second global oxygenation episode, GOE2, which occurred about 0.8 to 0.6 Ga (see Chapter 1). This event involved what turned out to be momentous increase in atmospheric oxygen concentrations, from levels of about 1% to 10–15% as shown in Fig 4.5. To summarize, there is increasingly robust evidence of a transition between 1.0 and 0.7 Ga whereby global photosynthesis went from being cyanobacterial-dominated to eukaryotic(algal)-dominated. This transition probably involved a significant relative decrease in cyanobacterial populations, coupled with

Fig. 5.3 The rise and fall of cyanobacterial stromatolites.

Stromatolites are layered rock formations formed by cyanobacteria over many thousands of years. This figure shows the abundance of cyanobacterial species in stromatolites over time. There is a striking peak during the latter half of the Proterozoic Eon, from about 1.6 to 1.2 Ga, followed by a steady decline. This decline corresponds to the period when algae were diversifying and is consistent with the algae gradually supplanting cyanobacteria as the major photosynthetic primary producers by the end of the Proterozoic.

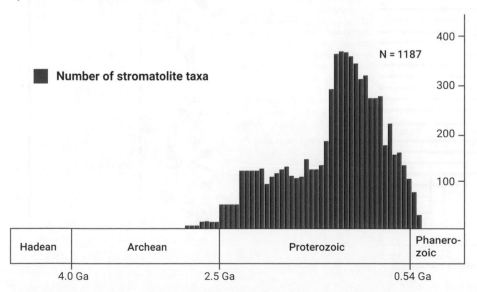

the rise of increasingly complex and diverse groups of multicellular green and red algae, such as *Bangiomorpha* and *Proterocladus*. It is estimated that in today's biosphere, cyanobacteria are only responsible for about a quarter of total photosynthetic primary production, despite containing over half of the global amount of chlorophyll. These figures clearly demonstrate the greater efficiency of eukaryotic versus prokaryotic organisms in terms of their photosynthetic productivity.

As levels of oxygen rose both in the atmosphere and in the oceans, the evolution of ever more complex multicellular aerobic organisms became possible (see Fig 5.4). By 0.8 to 0.7 Ga, early types of multicellular green algae, such as *Cladophora*, were present in many of the extensive oceans and shallow seas that covered nearly all of the Earth at this time. Between 0.6 and 0.4 Ga, there was an explosive growth of new forms of red, brown, and green algae. Brown algae are derived from non-algal cells that ingested smaller unicellular red algae and converted them into secondary

Fig. 5.4 Diversification of the primary algae from 1.0 Ga to now.

Phylogenetic relationships among the main green algal lineages leading to land plants. At the top are the three most common chlorophyte groups, Ulvophycheae, Chlorophyceae, and Trebouxiophycheae. At the bottom are the streptophytes including the algal charophytes and the land plants. Drawings illustrate representatives of each lineage: (1) *Acetabularia*, (2) *Pediastrum*, (3) *Chlorella*, (4) *Tetraselmis*, (5) *Picocystis*, (6) *Ostreococcus*, (7) *Micromonas*, (8) *Crustomastix*, (9) *Monomastix*, (10) *Pyramimonas*, (11) *Pycnococcus*, (12) *Pseudoscourfieldia*, (13) *Nephroselmis*, (14) *Prasinococcus*, (15) *Verdigellas* (a: general habit, b: individual cells in a gelatinous matrix), (16) *Mesostigma*, (17) *Chlorokybus*, (18) *Klebsormidium*, (19) *Chara*, (20) *Xanthidium*, (21) *Chaetosphaeridium*, (22) *Coleochaete*, (23) *Ranunculus*.

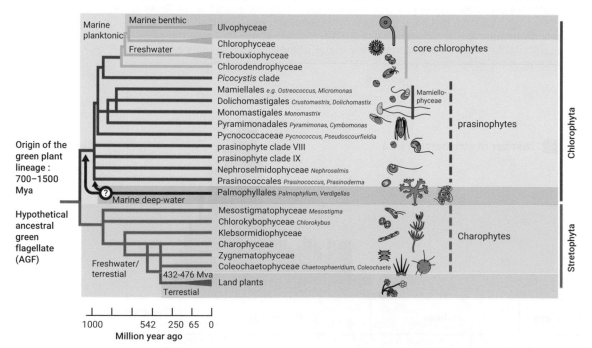

endosymbionts (see section 5.3 below). DNA analysis suggests that some brown algae subsequently lost their plastids but went on to acquire groups of fungal genes via horizontal gene transfer (HGT). Some of these chimeric species developed into superficially fungus-like organisms, although in reality they are probably unrelated to fungi, which are classified alongside metazoans in the Opisthokont group. One group that might be derived from these aplastidial brown algae is the oomycetes or water moulds, which include some of the most serious pathogens of plants, including many major crops such as potatoes (see section 5.4 below).

Within the green algae, some groups remained either unicellular or colonial while other lineages diversified into true multicellular organisms with some degree of tissue differentiation. Multicellular green algal forms included branched or unbranched filaments, and membranous or tubular thallus structures. Predation by single-celled eukaryotes might have led to the evolution of the first armoured algae, around 0.8 Ga. These armoured algae include some photosynthetic dinoflagellates that are encased in cellulosic plate-like structures called theca, which have a mechanical strength comparable to some of the softwoods of vascular plants. Soon after this, there was a major increase in overall eukaryotic diversity 'Ediacaran Biota' during the final stages of the Proterozoic Eon, between about 0.64 and 0.55 Ga, as discussed in Chapters 1 and 7. Further into the Phanerozoic Eon, after about 0.55 Ga, the Chlorophytes continued to play a dominant ecological role with a continual process of extinction and the appearance of new forms in response to environmental changes. For example, vast numbers of these algal species were lost during the mass extinction event at the Cretaceous/Tertiary boundary that also killed off the non-avian dinosaurs 66 Ma.

Today, only about 10% of the approximately 4,300 known Chlorophytes species are marine and the vast majority now occupy various freshwater niches, while several have formed symbiotic relationships with protozoa and fungi.

The Streptophyte lineage probably dates from about 1.2 to 1.0 Ga and includes a diverse array of unicellular and multicellular green algae that occupy freshwater and semi-terrestrial environments (collectively termed the Charophytes), plus the true land plants (Embryophytes). Phylogenomic analyses indicate that the earliest-diverging Streptophytes included morphologically simple unicellular species of the Mesostigmatophyceae, Chlorokybophyceae, and Klebsormidiophyceae with packets of non-motile cells and multicellular gametophytes in the latter (see Fig 5.5). Later developing Streptophyte groups evolved several significant innovations that pre-adapted them to more complex multicellularity and gave them the potential to move out of their aquatic habitats. For example, at the molecular level, innovations such as the evolution of auxin and abscisic acid signalling pathways were already underway in some unicellular Streptophytes as early as 1.0 Ga.

More advanced Streptophytes, such as the Charales or stoneworts, developed several innovations that ultimately led to the development of complex tissues and large plant bodies. Examples include filamentous thalli with

Fig. 5.5 Mechanism of secondary endosymbiosis.

Secondary endosymbiosis occurs when an ancestral host cell engulfs a photosynthetic eukaryotic alga. The alga already has a chloroplast with two membranes as well as a nucleus and other organelles. Since the host cell only needs the photosynthetic capacity of the algal chloroplast, the other captured organelles, including the nucleus, degenerate and eventually disappear. However, the membranes often remain and the chloroplast is initially left with four membranes, although in some secondary algae one or more of the 'extra' membranes are eventually lost.

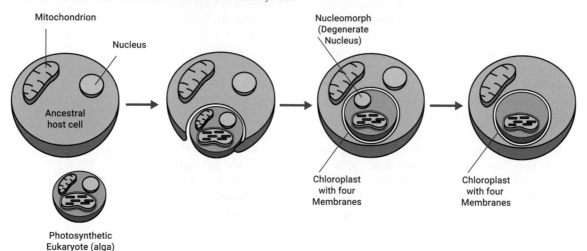

branching cell differentiation and apical growth, plus a new mechanism of cell division involving the production of a phragmoplast. This is the structure formed during cell division that develops into the new cell wall that separates the daughter cells. Also, Charophyte cell walls contained regulated channels called plasmodesmata that linked adjacent cells. These structures greatly facilitated intercellular communication and the efficient flow of materials, such as sugars and other photosynthetic assimilates, throughout the increasingly large and structurally differentiated algal body. In today's large land plants, such as trees, plasmodesmata can mediate the rapid transport of assimilates through the phloem network over the distances sometimes in excess of 100 metres. The Streptophytes adapted to life in freshwater habitats very early in their evolution, and were probably the first eukaryotic algae to colonize such non-saline ecosystems. These habitats were also adjacent to a new terrestrial environment, the land surface, that was largely free of competing organisms, with the probable exception of a few restricted communities of cyanobacteria and other prokaryotes.

Unlike most oceanic habitats, many of the freshwater habitats colonized by Streptophytes, such as rivers and smaller lakes, underwent frequent and often seasonally regular episodes of drying out followed by re-inundation. These environmental conditions provided an ideal suite of selection pressures that caused some of the increasingly complex Streptophytes, such as the *Charales*, to develop specific adaptations to a terrestrial existence including some measure of desiccation tolerance. By the end of the Proterozoic, about 0.55 Ga, a sister lineage to the Charales, probably more closely related to the extant Zygnematophyceae, succeeded in making the first decisive steps towards permanent colonization of the land surface. These organisms became the Embryophytes, or land plants. By successfully adapting their morphology, physiology, life history, and reproductive mechanisms to their new terrestrial habitats, the Embryophytes were able to undergo an explosive radiation after 0.55 Ga. The subsequent evolution and diversification of the Embryophytes to become the dominant global flora is explored further in Chapter 6.

5.3   Secondary and tertiary endosymbiosis

Primary endosymbioses are extremely rare evolutionary events and there are only two known examples (see Chapter 4). However, in addition to these rare primary endosymbioses, green and red algae have frequently participated in secondary and tertiary endosymbiotic events. These higher order endosymbioses have given rise to many plastid-containing organisms that superficially resemble green or red algae. A secondary endosymbiosis occurs when a heterotrophic host cell captures either an entire red or green algal cell that already contains its own cyanobacterial-derived plastid. The engulfed algal cell is subsequently relegated to the status of a plastid organelle. This process results in a new type of 'secondary' algal cell made up of the original eukaryotic heterotrophic host cell that contains complex plastid(s) derived from an engulfed eukaryotic alga (see Fig 5.5). These so-called 'complex' algae can be found in many diverse lineages across most of the major eukaryotic supergroups.

> The extremely small number of primary plastid acquisitions compared with the many dozens of eukaryotic lineages with complex plastids suggests that the capture and 'domestication' of a cyanobacterial endosymbiont *de novo* is a highly challenging process in evolutionary terms, whereas the capture/domestication of an algal symbiont is a lot more straightforward.

Primary endosymbiosis requires complex adaptations to fully integrate the cyanobacterium into its host cell metabolism and to convert it to a permanent organelle. The adaptations required for such a 'domestication' include the evolution of metabolite export, protein import, and redox regulation systems, plus transfer of plastid genes to the host nucleus and their conversion to eukaryotic genetic requirements. It appears, however, that once the primary plastid has already become fully integrated within a host cell, it becomes much easier for such algae and their descendants to be ingested and converted into secondary plastid endosymbionts by other eukaryotic hosts. This process can be taken further, for example if a complex alga formed as a result of a secondary endosymbiosis is in turn ingested by another heterotrophic cell, thereby creating a tertiary or even higher order endosymbiosis. In one remarkable example, the kryptoperidiniacean diatom group has possibly undergone as many as ten separate episodes of tertiary endosymbiosis!

In contrast with primary endosymbiosis, which only ever gave rise to descendants in the Viridiplantae, secondary and tertiary endosymbioses have led to photosynthetic descendants in at least six other major groups of protistan eukaryotes, namely Haptophytes, Alveolates, Stramenopiles, Rhizaria, Excavates, and Cryptophytes. In all of these cases the identity of the heterotrophic host cell is still unknown but invariably the ingested cell was either a red and/or green alga or a secondary derivative thereof. The widespread distribution of these complex endosymbioses across the eukaryotic tree of life is shown in Fig 5.6. The complex plastids in secondary algae are in reality the remains of entire algal cells and are typically surrounded by either three or four sets of envelope membranes, instead of the double envelope membranes found in the primary plastids of Archaeplastida. As they were converted into much-reduced plastid organelles, the engulfed red or green algal cells eventually lost most of their eukaryotic characteristics, such as mitochondria, endoplasmic reticulum, and flagella. However, in most cases their primary plastid(s) were retained and remained fully functional with regard to photosynthesis and other typical plastidial activities, such as fatty acid and starch biosynthesis. Many genes of either cyanobacterial or eukaryotic ancestry were transferred from the endosymbiont nucleus to the nucleus of the secondary host. The endosymbiont nucleus gradually became ever more reduced, and in most cases eventually disappeared.

An exception to this is found in the Cryptophyte and Chlorarachniophyte algae where the endosymbiont nucleus persists as a relict 'nucleomorph' that resides in the periplastidial compartment, which is derived from the cytosol of the engulfed alga. While the original primary plastids within the endosymbiont retain most of their genomes, the nucleomorphs have lost nearly all of their genes, with their genomes becoming reduced to tiny structures of about 0.5 Mb of DNA. It seems likely that this process of gene loss normally leads to complete loss of the nucleomorph since this structure is absent in all of the other groups of secondary plastid-bearing algae. The reason for the persistence of a nucleomorph in some algae has yet to be determined. One possibility is that some of its few remaining genes still have essential functions but cannot be spliced out for transfer to the new host nucleus and therefore the entire nucleomorph is maintained. In most secondary endosymbioses, the presence of a nucleomorph is an intermediate stage prior to its complete loss.

Complex endosymbioses mainly from green algae

Until recently, it was generally thought that green algae were engulfed as principal secondary endosymbionts only three times during evolution. This is because only three groups of complex algae have plastids that resemble green algae, especially in terms of their pigment compositions. In each of the three cases considered below, what appears to be a recognizable Chlorophyte green alga was taken up by a heterotrophic host cell belonging respectively to either the Excavate, Rhizaria, or Alveolate eukaryotic supergroups. However, as discussed later in this section, this view might need

Fig. 5.6 Eukaryotic tree of life showing plastid-bearing lineages and their closest relatives.

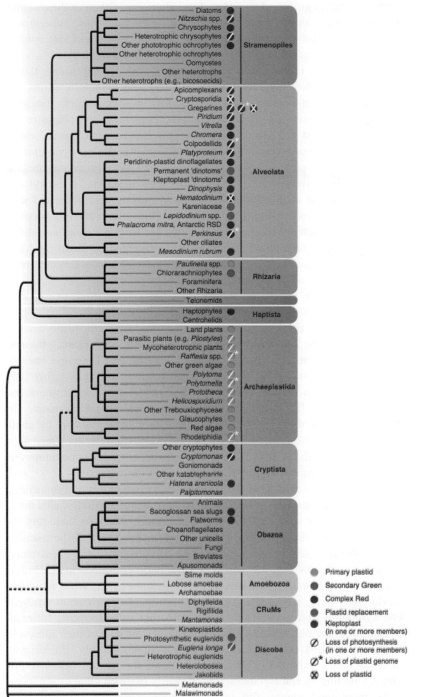

The tree topology is based on recent analyses from 2019 and 2020. The type of plastid (primary or secondary/complex) is indicated next to each lineage. Where known, specific complex events of kleptoplasty and plastid replacements (serial secondary or tertiary) are shown. Known instances of loss of photosynthesis are indicated with a line through the plastid circle; loss of photosynthesis with loss of the plastid genome is indicated by a line through the plastid circle and an asterisk. Complete loss of a plastid is indicated by two lines through their plastid circle. Dashed lines in the tree represent regions of uncertainty with respect to the phylogenetic placement of the corresponding lineages.

to be revised as more algal genome sequences are analysed and a picture is gradually emerging of possible multiple green and red endosymbiotic events in several groups of complex algae. Therefore, although the complex plastids that are currently present in these three eukaryotic supergroups often appear to be of green algal origin, the host cells might also contain cryptic remnants of other red and/or green algal endosymbionts.

The Excavates constitute a diverse supergroup of unicellular, flagellate eukaryotes that includes both free-living and symbiotic forms. All Excavates are non-photosynthetic except for the monophyletic Euglenid group of Euglenozoans. Although a few Euglenids are phagotrophic heterotrophs, most are photosynthetic, freshwater-adapted species with a green algal-like endosymbiont that is similar to free-living *Pyramimonas* spp. The complex plastids in Euglenids have three membranes and the host cells store the carbohydrate paramylon, instead of starch, in the cytosol. Excavates are one of the earliest diverging groups of eukaryotes and were probably present as free-living heterotrophs within 100 to 200 million years of LECA. The Euglenid endosymbiont is derived from a *Pyramimonas*-like Prasinophycean green alga that had possibly diverged by about by 1.2 to 1.4 Ga, so the Euglenids as a group of photosynthetic eukaryotes most likely date from around the middle of the Proterozoic Eon.

Similarly to the Excavates, the Rhizaria are a supergroup of unicellular eukaryotes that are amoeboid. They are all non-photosynthetic, with the exception of the Chlorarachniophytes. The latter constitute a small group of marine plastid-bearing photosynthetic organisms. Because Chlorarachniophytes are also motile and are capable of ingesting bacteria and small eukaryotic cells, they are regarded as mixotrophs rather than being obligate autotrophs. Chlorarachniophyte plastids have three membranes and store β-1-3-glucans, instead of starch, in their cytosol. The Rhizarian lineage diverged shortly after the Excavates, probably by about 1.6 Ga, while the likely source of the plastid, an Ulvophyte-like green alga, was present by about 1.0 to 1.2 Ga. Therefore, the Chlorarachniophytes might date from as early as 1.0 Ga.

The Alveolates are another structurally diverse supergroup of unicellular eukaryotes. They include a wide range of heterotrophic ciliates, plus many photosynthetic or secondarily non-photosynthetic protozoa including the highly abundant dinoflagellate group of phytoplankton and the parasitic Apicomplexans (see below, section 5.4). Dinoflagellates are extremely versatile organisms that are found in all aquatic environments including marine, brackish, and fresh water, as well as in snow or ice. While most species are free living, an important group of dinoflagellates are photosynthetic endosymbionts that reside inside corals and other animal species. Other dinoflagellates are parasitic on animals or protists for part of their life cycle, but have retained their plastids and photosynthetic ability. Unlike most other dinoflagellates, which harbour red algae, the species *Lepidodinium viride*, contains an endosymbiont that is clearly of green algal origin.

As mentioned above, the relatively straightforward story of three episodes of secondary endosymbiosis, involving a single green alga and a different heterotrophic host cell, has recently been challenged. This reappraisal is due to new genomic data that is continually emerging from the ever-increasing number and diversity of algal species that have been fully sequenced with respect to their nuclear and plastid genomes. It seems likely that the three plastids described above in their respective Excavate, Rhizaria, and Alveolate hosts, are derived mainly from green algal endosymbionts. Yet, in each case several additional genes of apparent red algal origin have now been found. Initially this was regarded as an artefact, but further analysis has revealed that in all three cases the nuclear genome is clearly a mosaic of genes from the original eukaryotic host that are mixed with additional genes of both green and red algal origin. Because there are more green algal genes than red algal genes, it is likely that the most important endosymbiont was a green alga. However, it appears that there were one or more prior or subsequent events that possibly involved the ingestion of a red algal cell and the transfer of some of its genes to the host genome before the rest of the alga was eventually digested.

A similar scenario that has been suggested for *Lepidodinium* is that it initially had a 'typical' dinoflagellate plastid of red algal origin, but then replaced it with a plastid from a green alga, while still retaining a few of the original red algal genes. In Fig 5.7, the chimeric nature of the genomes of the main supergroups of secondary photosynthetic eukaryotes is shown.

Fig. 5.7 Genes of red and green algal ancestry in secondary photosynthetic eukaryotes.

The number of red or green algal-like genes is shown in each lineage and is classified according to their origin and statistical support in phylogenetic trees. This shows that in each of the major lineages of secondary algae their genomes contain a mixture of genes derived from both red and green primary algae. As discussed in the main text, it is likely that these genes are derived from multiple serial endosymbiotic events.

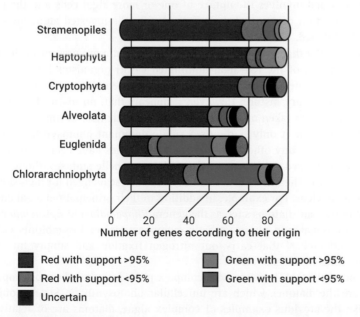

Note that in each supergroup there is a mixture of genes of green and red algal origin but, as discussed in the next section, red algal genes greatly outnumber green algal genes in the CASH supergroup (see below). These new discoveries highlight the apparent frequency of complex endosymbioses, and the evident adaptive advantages that can accrue for a diverse range of eukaryotic heterotrophs if they are able to capture and exploit a fully functioning photosynthetic eukaryotic cell. Indeed, even after the heterotrophic host has 'captured' an algal cell and converted it into an endosymbiont, it is not precluded from subsequently ingesting one or more additional algal cells to serve as additional or replacement endosymbionts. In such cases, it seems that only one endosymbiont is fully retained in the long term, and the other(s) are eventually digested after any useful genes are salvaged from their genomes and transferred to the host cell nucleus.

Complex endosymbioses mainly from red algae

As discussed in the previous section, complex endosymbioses can involve multiple episodes of the capture of red and/or green algal cells and their conversion into plastids. The Cryptophyte, Alveolate, Stramenopile, and Haptophyte (CASH) lineage encompasses four diverse and ecologically important eukaryotic supergroups that include both photosynthetic and non-photosynthetic groups. It was originally thought that all of the complex red plastids in the CASH lineage were descended from a single episode involving the ingestion of a red algal cell. As more sequence data have become available, this is now thought increasingly unlikely. Instead, it is proposed that complex red plastids were acquired via several independent events including serial tertiary and higher order endosymbioses. In addition, kleptoplasty is now more frequently observed in a range of different host cells and involves the uptake of one or more algal cells and the possible extraction of some of their genes before the ingested algae are ultimately digested.

The finding that some genes of green algal origin are present in the nuclear genomes of the red-algal-dominated CASH organisms (see Fig 5.8), indicates that the host cells had also ingested green algae at some point in their evolutionary history. However, it appears that, no matter how many algae have been taken up by a complex eukaryotic host during its history, at any one time it only harbours a single dominant photosynthetic plastid. Presumably any other ingested algae or plastids have been either completely eliminated or downgraded to relict organelle status, although not before some of their more useful genes have been scavenged for inclusion in the host nucleus. For example, in addition to their principal red algal endosymbiont, some diatoms such as the genera *Rhopalodia* and *Epithemia* contain additional non-photosynthetic cyanobacterial endosymbionts called 'spheroid bodies' that carry out nitrogen fixation and supply nitrogen compounds to the host cell.

Among the CASH groups with complex plastids, by far the most important are the diatoms, which are unicellular photosynthetic Stramenopiles. Unlike the previous examples of complex algae, diatoms are of relatively recent origin. They are probably originally derived from at least two serial

Fig. 5.8 Giant kelp forest off the coast of Tasmania, Australia.

The canopy of this 'forest' is dominated by the giant kelp, *Macrocystis pyrifera*, which is a secondary brown alga. The main generation of this macroalga is a perennial diploid sporophyte that can grow up to 45 m in length. Such kelp forests harbour species-rich ecosystems that include other algae and numerous fish and invertebrates, as well as diving birds and mammals such as seals, sea lions and sea otters. Human activities are now threatening many kelp forests around the world.

Source: C. L. Butler, V. L. Lucieer, S. J. Wotherspoon, C. R. Johnson, 'Multi-decadal decline in cover of giant kelpMacrocystis pyriferaat the southern limit of its Australian range' in Marine Ecology Progress Series, 653: 1-18, 2020, fig. 1. Licensed under CC BY 4.0. DOI: 10.3354/meps13510 [link source URL to DOI]

endosymbioses that occurred between 1.2 and 0.7 Ga, although diatoms did not radiate in their present calcified form until after 0.2 Ga. Today, the calcified diatoms are one of the most ecologically important groups of photosynthetic organisms, contributing to as much as 30–40% of global primary production in some regions at certain times of the year when supplies of nutrients and sunlight are optimal. Diatoms are also responsible for as much as 20% of global oxygen evolution although, because this only occurs in the aquatic habitats, much of it is rapidly consumed by other aquatic organisms rather being released into the atmosphere. Due to their rapid growth rates and highly scalable productivity, diatoms are now being assessed for the commercial cultivation of a wide range of renewable products from biofuels to specialty chemicals.

Diatoms remained as minor components of global marine ecosystems until as recently as 0.1 Ga when the breakup of the supercontinent, Pangea, led to a great expansion in the oceans and an increased delivery of nutrients from the drifting continental land masses. Sequencing studies reveal that diatoms have chimeric nuclear genomes similar to those seen in other products of complex endosymbioses as discussed above. Hence, similarly to the dinoflagellate, *Lepidodinium* (see previous section), diatom genomes contain genes derived from both red and green algae, as well as numerous genes of bacterial origin. Although it is generally thought that the original endosymbiont was a red alga, the genomic data suggest that both red and green algae have contributed significantly to extant diatom nuclear genomes to such a degree that it is not possible to say conclusively which type of alga came first. The presence of bacterial genes suggests that the heterotrophic ancestors of diatoms also fed on a range of bacterial prey, including proteobacteria, rhizobia, and cyanobacteria, from which they were able to acquire useful genes. These bacterial genes enabled diatoms to acquire novel metabolic capabilities including the ability to construct their unique silicon-based cell walls, or frustules.

At the opposite end of the size spectrum to the tiny diatoms is a related Stramenopile algal lineage, the brown algae or *Phaeophyceae*, so-called due to their greenish-brown fucoxanthin pigments. Brown algae are all multicellular and probably diverged from flagellate unicellular algal ancestors between 0.2 to 0.3 Ga. The original endosymbiont was probably a red alga from which cellulose synthase genes were acquired, while their alginate pathway genes were probably obtained via HGT from an actinobacterium. Alginates from brown algae have many uses, such as thickening agents and stabilizers in foods, or as gels in textile and medical applications. Brown algae are of particular evolutionary interest because they are one of only four major biological lineages that have developed complex multicellular organizations—the others being the Viridiplantae, Fungi, and Metazoa. One of the best-known groups of brown algae are the giant kelps, such as *Macrocystis*, which can grow to over 45 metres in length (see Fig 5.8). These kelps are an important, often dominant, component in many coastal ecosystems where they typically grow in communities, known as kelp 'forests', that harbour a striking diversity of other marine species, including sea otters. Despite their large size, kelps and other brown algae are significantly less complex than land plants in terms of their tissue organization and physiological flexibility and are largely restricted to coastal habitats where they continue provide important, albeit localized, ecosystem services.

5.4   Loss of photosynthesis and plastids leads to new types of organisms

The acquisition of oxygenic photosynthesis has resulted in enormous evolutionary success for most organisms, enabling them to dominate global ecology for billions of years. This makes all the more surprising that photosynthesis has been either partially or completely lost in many eukaryotic groups.

Some of the many examples of the loss of plastid genomes or even entire plastids are shown in Fig 5.7. It seems that in such cases the development of alternative non-photosynthetic lifestyles, often involving parasitism, were more evolutionarily adaptive options for the species concerned. As discussed below, this phenomenon is particularly marked in eukaryotic species that have undergone secondary endosymbioses. As discussed in Case study 5.2, it is now apparent that there are many examples of partially

Case study 5.2

How photosynthesis was repeatedly gained and lost in the algae

Gaining photosynthesis

There are two ways for a unicellular, heterotrophic eukaryote to become photosynthetic, namely, either ingest a cyanobacterium and convert it into a simple primary plastid, or alternatively to ingest another photosynthetic eukaryote (ie an alga) and convert it into a complex secondary plastid. It might seem logical that the first route to photosynthesis, which involves capturing and 'domesticating' a small and simple bacterial cell, would be much easier than the second route, which involves capturing a relatively large and complex eukaryotic cell. However, there is now good evidence that the first route of primary endosymbiosis is extremely rare and difficult, while secondary endosymbiosis is much more common and easier. It appears that the most difficult step in acquiring photosynthesis is the 'taming' of a free-living cyanobacterium and its conversion into a fully functional plastid organelle, which as far as we know only seems to have happened twice in evolution.

In contrast, there have been numerous occasions when a heterotrophic protist engulfed a red or green alga and converted the entire algal cell into a complex plastid. The resultant secondary algae are extremely diverse and have successfully evolved several different lifestyles. Some of them have remained motile as facultative unicellular heterotrophs, while others have developed into complex multicellular plant-like organisms, such as the giant kelps that form coastal ecosystems similar to terrestrial forests (see Fig 5.8). Puzzlingly, however, many secondary algae went on to lose this seemingly highly advantageous trait.

Losing photosynthesis

It might seem obvious that the ability to photosynthesize should be of great benefit to an organism. Instead of relying on feeding upon other living or dead organisms as heterotrophs must do, a photosynthetic autotroph can use energy from sunlight to synthesize its own complex organic compounds. So why, as discussed in the main text, have so many algae partially or completely lost the ability to photosynthesize? The answer is that the lifestyle change must involve a net evolutionary advantage that outweighs the disadvantages of losing photosynthesis, otherwise the organism would not be competitive in terms of natural selection. Although photosynthesis has many advantages in terms of not requiring organic carbon sources, some external resources such as light and mineral nutrients are still required and access to these can be highly competitive.

In the oceans, suitable amounts of light are restricted to near-surface habitats whereas nutrients might be located at lower depths. Also, many photosynthetic algae have retained some degree of motility meaning that they can readily revert to an active 'hunting' lifestyle. Indeed, there are examples of mixotrophic algae that can switch between hetero- and auto-trophic lifestyles depending on the relative availability of light, nutrients, and food sources. In other cases, such as in corals, an alga might have originally been a photosynthetic symbiont that developed into a parasite. By losing its photosynthetic apparatus, the alga required less energy and it was able to obtain this from its former symbiotic partner without killing it, thereby establishing a more durable form of parasitism. As ever, there is a cost-benefit relationship in any evolutionary strategy and the benefits of photosynthesis can sometimes be outweighed by its costs.

or totally non-photosynthetic eukaryotes that were originally fully photosynthetic, obligate autotrophs.

In some cases, these eukaryotes have retained some photosynthetic capacities, while mainly operating as facultative heterotrophs. In the majority of cases, however, photosynthesis, and sometimes part or all of their original plastid organelles, have been lost so that the organisms are now obligate heterotrophs. Some of these heterotrophs are free-living, but many have become obligate parasites that are totally dependent for their survival on a host organism. Indeed, several descendants of formerly photosynthetic organisms have now evolved into some of the most important medical and agricultural pathogens that afflict humans and their crop plants. These include several protozoan parasites responsible for major human infectious diseases including malaria, cryptosporidiosis and toxoplasmosis. Other possible examples include the oomycete pathogens responsible for devastating crop diseases, such as potato blight and soybean rot. Several examples of the evolution and impact of these secondarily non-photosynthetic eukaryotes will now be discussed.

Archaeplastida

There are numerous examples of the partial or total loss of photosynthesis in the Archaeplastida, both in the aquatic green algae and in land plants. Examples of a deviation from obligate autotrophy can be found in unicellular algae that have become facultative heterotrophs. Such species are normally found in nutrient-limited or light-limited environments where the primarily autotrophic lifestyle of the alga is supplemented by periodic opportunistic predation when circumstances demand it. Such algae are still fully competent as photosynthetic autotrophs and have not lost any genes related to these functions. They can therefore be regarded as having a mixotrophic lifestyle. In other cases, the plastids have ceased to be functional. Such an example of complete reversion to a non-photosynthetic state in the Archaeplastida is found in two *Rhodelphis* species (see Fig 5.9). These organisms constitute a recently characterized sister group to the red algae that probably diverged from

Fig. 5.9 *Rhodelphis*, a red algal relative that has lost its photosynthetic ability.

This formerly photosynthetic red alga has now reverted to a heterotrophic lifestyle but has retained its plastid, which has acquired new functions. Recent genome and transcriptome analyses reveal that the *Rhodelphis* plastid now contains proteins with FeS clusters that probably cooperate with mitochondria in haem biosynthesis. Isoprenoid and fatty acid biosynthesis have moved to the cytosol. The cells are capable of phagocytosis for the capture of prey and contain two flagella for motility. Note that subcellular structures are not drawn to scale and the mitochondrial genome has not been verified to be circular.

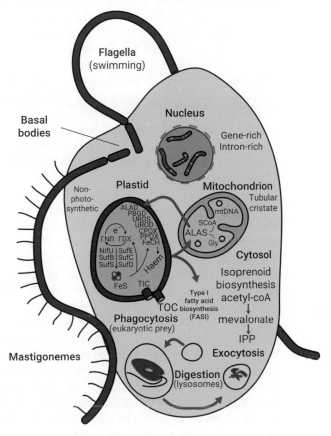

their photosynthetic relatives soon after the first red algae appeared about 1.6 to 1.8 Ga. *Rhodelphis* spp. are phagotrophic, flagellate predators with much-reduced or absent plastids.

Extant *Rhodelphis* spp. have now lost almost all photosynthetically related genes, including those of light-harvesting, electron transport, and pigment biosynthesis. An interesting feature of *Rhodelphis* is that fatty acid biosynthesis, which normally occurs in plastids using multi-subunit type II fatty acid synthase, now occurs in the cytosol and is mediated via a multi-domain type I fatty acid synthase. This situation is unique in the Archaeplastida, where all other known species use a plastid-located, cyanobacterial-derived, multi-subunit type II fatty acid synthase. It appears that, although *Rhodelphis* spp. have lost their original cyanobacterial type II fatty acid synthase genes along with all the other plastid functions, they have been able to compensate for this otherwise fatal loss by acquiring a type I fatty acid synthase from elsewhere. The genes encoding this enzyme could have been acquired via HGT from an animal, fungal, or protist donor genome that was ingested as part of their predatory lifestyle.

There are also two non-photosynthetic genera in the Trebouxophyte green algae, namely *Prototheca* and *Helicosporidium*, both of which are free living heterotrophs found in soil and aquatic ecosystems. These algae are closely related to the two well-known photosynthetic genera, *Chlorella* and *Auxenochlorella*. Similarly to the red alga, *Rhodelphis*, discussed earlier in this section, all of the non-photosynthetic green algae have retained a colourless plastid with a reduced genome. Although all of the genes directly related to photosynthesis have been lost in these algae, they have retained several genes of cyanobacterial origin that are related to other vital functions such as fatty acid and terpenoid biosynthesis. Total loss of the plastid genome in non-photosynthetic primary algae is extremely rare with only one known possible example in the *Polytomella* genus, which are relatives of the much-studied model green alga *Chlamydomonas reinhardtii*.

Stramenopiles

The Stramenopiles are part of the highly diverse CASH lineage (see above) where loss of photosynthesis is relatively common. Stramenopiles are characterized by possessing two unequally sized flagella (hence their alternative name, Heterokonts). Non-photosynthetic and plastid-lacking Stramenopile groups include oomycetes such as *Phytophthora*, the causative agent of potato blight, and *Blastocystis*, an intestinal parasite of humans. Oomycetes have many fungal-like characteristics and were originally believed to be a divergent group of fungi. However, genomic sequencing studies show that they are clearly derived from non-fungal ancestors. Phylogenetic analyses show that oomycetes share a common ancestor with the diatoms and brown algae, raising the possibility that they are derived from photosynthetic algae. Unlike fungi, which have chitin-containing cell walls, oomycetes have cellulose-based cell walls that are similar to those of plants and algae. This has led to suggestions that oomycetes were originally marine organisms that possibly diverged from a diatom-like photosynthetic ancestor during the late Proterozoic, about 0.6 to 0.8 Ga, although there is some uncertainty about the timing and mechanism of their origins. Also, if they once had plastids, they no longer have any remnants of these organelles. The earliest oomycete fossils are dated at about 0.4 Ga and molecular clock estimates place their origins slightly before that time.

At some point, oomycetes became parasitic, a lifestyle that meant they were in intimate contact with their various host species, which included many fungi and algae. This allowed the oomycetes to acquire genes from their various hosts via HGT. For example, present day oomycetes contain numerous genes derived from ascomycete fungi. These genes, which include lipid biosynthesis components, have enabled oomycetes to develop a filamentous growth habit that is strikingly similar to fungi. Indeed, genome analysis shows that oomycetes are promiscuous genetic chimeras with genes acquired from many sources including various algae, fungi, and bacteria. The impressive ability of oomycetes to acquire and usefully deploy large numbers of genes from unrelated organisms has led to some questions about whether they really are descended from photosynthetic diatom-like ancestors or simply acquired their algal-like genes via HGT. Therefore, although many phylogenetic analyses place oomycetes as very close relatives to diatoms, with which they may share a common ancestor, the full details of their convoluted evolutionary history have yet to be determined conclusively.

Alveolates

Like Stramenopiles, the Alveolates are a diverse supergroup within the CASH lineage. They are characterized by the presence of a continuous layer of flattened vesicles (alveoli) immediately inside their cell membrane. Alveolates include major photosynthetic groups, such as dinoflagellates and chromerids. Another large Alveolate phylum is the Apicomplexans, which is mostly made up of unicellular, spore-forming, obligate parasites of vertebrate animals. Apicomplexan cells typically contain a much-reduced plastid-like organelle, called an apicoplast, that is surrounded by four envelope membranes and is probably derived from a red alga. Apicoplasts have lost many of their original plastidial functions, including all activities related to photosynthesis. Nevertheless, they are still the sites of several important metabolic pathways found in conventional plastids, including fatty acid and isoprene biosynthesis. Removal of apicoplast from its host cell results in the loss of the ability of the parasite to invade new host cells and its eventual death. Therefore, despite its reduced status in terms of size and complexity, the apicoplast remains essential for parasite function, which makes it a useful target for drugs aimed at disrupting its life cycle, as we will now see.

Most extant Apicomplexans are parasites of animals and they include several serious pathogens of humans and other animals. Genomic data suggest that free-living Apicomplexans were already present by 0.6 to 0.8 Ga, which implies that their original hosts were most likely marine invertebrates, because terrestrial vertebrates had not evolved until after 0.4 Ga. Probably the best-known Apicomplexans in terms of their impact on humans are *Plasmodium falciparum*, the causative agent of malaria (see Fig 5.10), and *Toxoplasma gondii*, which is estimated to infect as much as 80% of the human population, albeit mostly in a latent and largely asymptomatic form. The relict plastid, or apicoplast, of *Plasmodium* is of considerable interest as part of efforts to develop more effective treatments for malaria. One of the few metabolic functions retained by this organelle is the biosynthesis of isoprenoids via the methylerythritol phosphate (MEP) pathway. The MEP pathway is essential for effective function of the malaria parasite but is not present in human hosts. For this reason, drugs that inhibit MEP pathway enzymes, such as fosmidomycin, are currently being developed as part of worldwide antimalarial strategies.

While most Apicomplexans contain apicoplasts, these relict organelles have been lost in some groups, such as *Cryptosporidium* spp., which is a well-studied human respiratory tract and gastrointestinal parasite. Infections by *Cryptosporidium* spp. are generally not very severe in healthy, immunocompetent individuals, but cause a respiratory and diarrhoeal disease that can be fatal in patients who are immunocompromised. The parasites are excreted from affected hosts and are often spread further via

The algal descendent, *P. falciparum*, is the most deadly species of *Plasmodium* that is responsible for mosquito-borne malaria in humans. The organelle labelled as an apicoplast is a highly reduced nonphotosynthetic plastid, left over from the original red alga from which *P. falciparum* is descended. The apicoplast has lost all photosynthetic genes and one of its few remaining functions is the biosynthesis of isoprenoids. The rhoptries, micronemes, and dense granules are secretory organelles at the apical pole, which enable the parasite to invade human red blood cells, leading to their lysis and the wider symptoms of malaria.

Fig. 5.10 *Plasmodium falciparum* **merozoite – a former alga now a malarial parasite.**

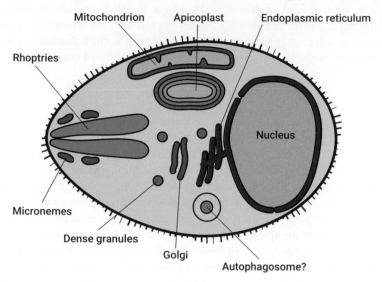

untreated or poorly treated water sources. Although they lack apicoplast organelles, *Cryptosporidium* spp. have several algal and cyanobacterial-like genes in their nuclear genomes. Examples include genes encoding glucose-6-phosphate and leucine aminopeptidase. The presence of these genes suggests that the ancestors of *Cryptosporidium* spp. once harboured red algal-derived plastids that were subsequently lost. Definitive evidence for the red algal origin of apicoplasts came from the discovery of several photosynthetic ancestors of Apicomplexans that were intracellular parasites of several corals. The parasitic Apicomplexans, informally termed corallicolids, live within the gastric cavity of their coral host cells. The parasite cells contain a single, large cone-shaped relict plastid bounded by four membranes that is structurally similar to the apicoplasts described above except that its genome has retained many more genes that are similar to those of red algal plastids. Although the apicoplast is no longer used for photosynthesis, it remains a highly active organelle that is the site of several essential cellular processes, including isoprenoid and heme biosynthesis.

In this chapter, we have seen how the original primary endosymbiotic event between a eukaryotic heterotroph and a cyanobacterium led to multiple lineages of photosynthetic organisms that ultimately included all of the main eukaryotic groups. The original primary endosymbiosis only led directly to a single daughter lineage, the Archaeplastida. From this group all extant primary red and green algae and the land plants emerged. The evolution of one group of green algae into land plants and their subsequent global radiation will be discussed in Chapter 6. Meanwhile, a wide array of other photosynthetic lineages was created from aquatic red and green algae via an additional series of secondary and tertiary endosymbioses. These involved eukaryotic heterotrophs from diverse backgrounds that ingested red and/or green algal cells and converted them into so-called complex plastids. Finally, there are many examples where both primary and secondary algae have subsequently lost their photosynthetic capacity and reverted to heterotrophic lifestyles, often as parasites.

 ## Chapter summary

- The direct descendants of the original primary endosymbiotic event at about 2.0 Ga are the Archaeplastida, which includes the red and green algae and land plants.
- The early red and green algae were highly efficient photosynthetic eukaryotes that exploited a wide range of marine habitats. These algae benefited from increased nutrient availability and by about 1.0 Ga their rates of primary production and oxygen evolution had overtaken those of the cyanobacteria.
- By about 0.6 Ga, oxygenic photosynthesis had increased atmospheric oxygen levels sufficiently to enable the evolution of complex multicellular plants and animals that set the stage for colonization of the land.
- After about 1.5 Ga, several new groups of eukaryotes acquired photosynthesis by engulfing red and green algae to form secondary endosymbioses. This created many secondary algal lineages including some key contributors to global photosynthesis.
- In contrast, some algal groups evolved into non-photosynthetic heterotrophs, including several important parasitic species of plants and animals.

 ## Further reading

Blaby-Haas CE, Merchant SS (2019) Comparative and functional algal genomics, *Ann Rev Plant Biol* 70, 1–23.34. DOI: 10.1146/annurev-arplant-050718- 095841
Overview of algal evolution and diversity in the light of new genome sequencing data.

Burki F, Roger AJ, Brown MW, Simpson AGB (2020) The new tree of eukaryotes, *Trends Ecol Evol.* DOI: 10.1016/j.tree.2019.08.008
New phylogeny of eukaryotes sheds light on the huge diversity of secondary algal groups.

Gawryluk RMR, Tikhonenkov DV, Hehenberger E, Husnik F, Mylnikov AP, Keeling PJ (2019) Non-photosynthetic predators are sister to red algae, *Nature* 572, 240–243. DOI: 10.1038/s41586-019-1398-6
Shows how some red algae quickly lost photosynthesis and became successful predators.

Gibson TM et al (2017) Precise age of *Bangiomorpha pubescens* dates the origin of eukaryotic photosynthesis, *Geology* 46, 135–138. DOI:10.1016/j.cub.2016.11.056
Analysis of a multicellular red algal fossil dates its appearance to about 1.0 Ga.

Oborník M (2019) Endosymbiotic evolution of algae, secondary heterotrophy and parasitism, *Biomolecules* 9, 266. DOI:10.3390/biom9070266
Good review of algal evolution showing how and why some algae became non-photosynthetic.

Suzuki S, Endoh R, Manabe R, Ohkuma M, Hirakawa Y (2018) Multiple losses of photosynthesis and convergent reductive genome evolution in the colourless green algae, *Prototheca*, *Scientific Reports* 8, 940. DOI:10.1038/s41598-017-18378-8
Useful case study of how several members of a green algal group independently lost photosynthesis.

Tang Q, Pang K, Yuan X, Xiao S (2020) A one-billion-year-old multicellular chlorophyte, *Nature Ecol Evol*. DOI: 10.1038/s41559-020-1122-9
Recent evidence that green algae became multicellular much earlier than previously thought.

Uitz J, Claustre H, Gentili B, Stramski D (2010) Phytoplankton class-specific primary production in the world's oceans: Seasonal and interannual variability from satellite observations, *Global Biogeochemical Cycles* 24, GB3016. DOI:10.1029/2009GB003680
Satellite data showing that global primary production is higher in algae than cyanobacteria.

Wang Q, Sun H, Huang J (2017) Re-analyses of 'Algal' genes suggest a complex evolutionary history of oomycetes, *Front Plant Sci* 8, 1540. DOI: 10.3389/fpls.2017.01540
Interesting exploration of the origin of 'algal' genes in oomycetes.

Discussion questions

5.1 Describe how red and green algae diversified and how some of them eventually developed into multicellular organisms.

5.2 What are the major differences between primary and secondary algae?

5.3 What have been the most important global impacts of algal photosynthesis?

5.4 Why do you think that many algae lost the ability to photosynthesize, and was this a good idea for them?

6 EVOLUTION OF THE LAND PLANTS

Learning objectives

- Understanding the global significance of land plant evolution.

- Examining the colonization of the land surface as a lengthy process where one group of green algae gradually evolved into the first terrestrial plants.

- Appreciating the many innovations required as the early **bryophytes** evolved into vascular plants that were better adapted to most terrestrial climates.

- Describing how pteridophytes and gymnosperms dominated the land flora for about 300 million years.

- Understanding the reasons behind the late diversification of the flowering plants, followed by their rise to becoming the dominant flora today.

- Learning how some plants partially or totally lost the ability to photosynthesize as they became parasites or pathogens.

6.1 Introduction

In this chapter we will examine the evolution of photosynthetic organisms on land. For the first three billion years since it evolved, the vast majority of photosynthetic life (and other life) was confined to aquatic habitats, mainly in the oceans. This immense timescale encompassed most of the Archean and Proterozoic Eons, from 4.0 **Ga** (billion years ago) to 550 **Ma** (million years ago). Although a few hardy cyanobacteria, algae and fungi had probably colonized a few restricted parts of the Earth's surface prior to 550 Ma, the land remained a mostly hostile habitat for life. This was especially true for the many complex multicellular eukaryotes, both photosynthetic and heterotrophic, that had evolved as part of the so-called 'Ediacaran Biota' during the final stages of the Proterozoic, between 635 and 550 Ma. The Ediacaran was one of the two periods during land plant evolution when there was a widespread burst of genomic novelty (the other was during the

Ordovician as discussed below). Genomic novelty was common in the streptophyte algae that were ancestors of land plants. These algae also developed increasingly complex multicellular forms with a wider range of tissue and organ types.

As discussed in Chapter 5, the increase in eukaryotic complexity during the Ediacaran, which also applied to ancestral groups of metazoans, coincided with the GOE2. The GOE2 led to oxygen to levels comparable with those of today and the establishment of a more effective and stable ozone layer. These factors created a completely new set of conditions that were favourable to the colonization of the land by complex eukaryotic life forms during the Cambrian Period, between 540 and 485 Ma. Around 470 Ma, during the mid-Ordovician Period, there is fossil evidence of spores from the first true land plants, the bryophytes, such as mosses and liverworts. By the late-Ordovician at about 450 Ma, putative spores from the more advanced vascular plants, such as ferns, have been found. This suggests that land plants may have already evolved into truly land-adapted organisms long before large terrestrial heterotrophs, such as animals and complex fungi established themselves on the land.

6.2   The importance of land plants

> The true land plants, or embryophytes, range from relatively small and simple non-vascular species, such as mosses and liverworts, to much larger and more complex vascular groups, such as ferns and the seed plants, or spermatophytes.

A broad overview of the evolution of the major land plant groups is shown in Fig 6.1. The seed plants, which include gymnosperms and angiosperms, consist of about 450,000 to 500,000 known species, although there are probably many more as yet undiscovered species and the true number could well be in the millions. These so-called 'higher plants' dominate most terrestrial ecosystems and have been popularly referred to as 'the lungs of the earth'. This is because of their huge emissions of oxygen gas as a by-product of photosynthesis. An arguably equally important role of these plants is fixing vast amounts of the important greenhouse gas, CO_2, thereby helping to avoid a potentially catastrophic increase in global temperatures. As we have seen in previous chapters, the oxygenated atmosphere that we now enjoy was originally created by relatively simple cyanobacteria many millions of years before land plants arose. After 2.0 Ga, this oxygenation process was continued and extended during the rise of photosynthetic eukaryotes—the algae—in aquatic ecosystems. However, as we will see in this chapter, land plants are much more efficient than marine organisms in terms of their overall photosynthetic primary production.

According to satellite measurements of chlorophyll fluorescence and other markers of active photosynthesis, as shown in Fig 6.2, it is estimated that global photosynthetic primary production, expressed as the amount of carbon (C) fixed per year, is relatively evenly balanced between terrestrial and oceanic habitats. Hence, the total global primary production is 104.9 Gt

Fig. 6.1 Evolution of the major land plant groups.

In this recent hypothesis, it is proposed that land plants are a monophyletic group descended from an ancestor called a stomatophyte. These organisms were characterized by the presence of stomata, as discussed in the main text. One group of stomatophytes diverged into the monophyletic bryophyte group (shaded green) in which the hornworts kept their stomata, while the mosses and liverworts lost them. The bryophytes were the first major group of land plants and were the dominant flora from about 430 to 360 Ma. A second group of stomatophytes developed into another monophyletic group, the tracheophytes or vascular plants. The tracheophytes became the dominant flora after 360 Ma and remain so to this day.

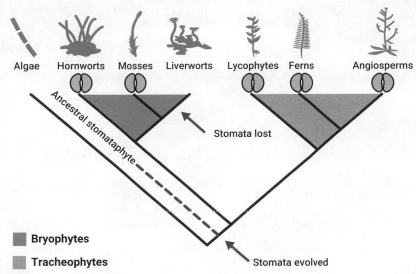

Source: B. J. Harris, C. J. Harrison, A. M. Hetherington, T. A. Williams, 'Phylogenomic Evidence for the Monophyly of Bryophytes and the Reductive Evolution of Stomata' in Current Biology, Cell, 30(11), 2001–2012, 2020, Graphical Abstract. Licensed under CC BY 4.0. DOI: 10.1016/j.cub.2020.03.048

C yr⁻¹ (Gt = giga or billion tonnes), of which 53.8% (or 56.4 Gt C yr⁻¹) is of terrestrial origin, while 46.2% of (or 48.5 Gt C yr⁻¹) is oceanic. However, this apparent near-parity is misleading because both the area and volume of the oceans are significantly greater than those of the combined terrestrial ecosystems. If we exclude the regions of permanent ice cover, where almost no photosynthesis occurs, the estimated average terrestrial primary production is 426 g C m⁻² yr⁻¹, while that of the much larger oceanic ecosystems is only 140 g C m⁻² yr⁻¹. This means that land-based photosynthesis is three-fold more productive than aquatic photosynthesis on an area basis, and this ratio is even higher on a volume basis.

Another significant difference between the land and the oceans lies in their standing stocks of biomass. Hence, although they are responsible for almost half of global photosynthetic production, oceanic autotrophs only account for about 0.2% of total biomass. This is because land plants sequester carbon into many additional long-lived compounds and structures that aquatic photosynthesizers do not possess. Notable examples include such highly durable compounds as lignin and other polyphenol derivatives and extracellular structures such as woody or cork-containing stems and roots. In many of the larger arboreal land plants the amount of 'dead' extracellular carbon, for example in tree trunks and branches, is

Fig. 6.2 Abundance of oceanic and terrestrial photosynthetic life.

This map is based on data generated from satellite images taken from September 1997 to August 2000. Note that the terrestrial flora (green) is largely concentrated in the tropics and northern temperate regions but is very sparse in the extensive semi-arid and desert regions (brown). The oceanic flora (pale green/yellow/red) is largely concentrated in coastal areas, including a surprisingly high abundance in the cool Antarctic waters where the low temperatures are mitigated by high levels of upwelling nutrients.

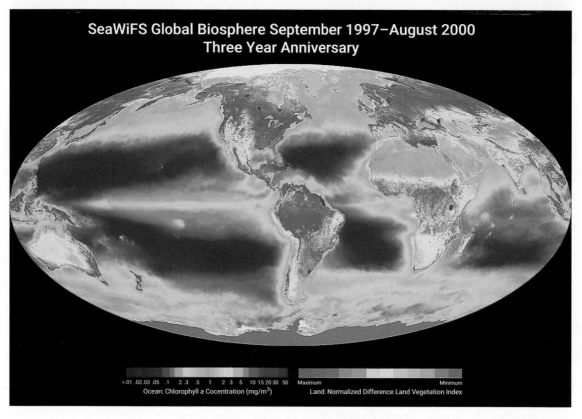

Provided by the SeaWiFS Project, Goddard Space Flight Center and ORBIMAGE

more than double that all of the 'living' intracellular carbon in the plant. In contrast, the carbon in most marine photosynthesizers is nearly all intracellular and tends to be relatively ephemeral. In general, eukaryotic photosynthesizers, such as diatoms and primary green algae, are much more efficient than cyanobacteria in terms of photosynthetic primary production.

> The evolution of land plants resulted in a further increase in overall photosynthetic productivity compared to their algal ancestors, which in turn greatly increased the total biomass and diversity of life on Earth.

During the Cambrian and early Ordovician Periods the first plants established themselves as semi-terrestrial organisms but remained confined in close proximity to water sources. During the later Ordovician and into Silurian Period, more truly land-adapted, or terrestrialized, plants evolved

Fig. 6.3 Changes in atmospheric O₂ levels from 600 Ma until today.

This shows how high levels of atmospheric O_2 (about 18%) were already established during the Cambrian (570–500 Ma). These levels increased during the early stages of land plant evolution during the Devonian (410–360 Ma), when non-vascular bryophytes dominated the terrestrial flora. This was followed by a huge increase to >30% during the pteridophyte-dominated Carboniferous from about 360–270 Ma. After this there was a sharp reduction to <15% that heralded the age of the seed plants, which are still the dominant vegetation, followed by a steady increase up to the current levels of 21%.

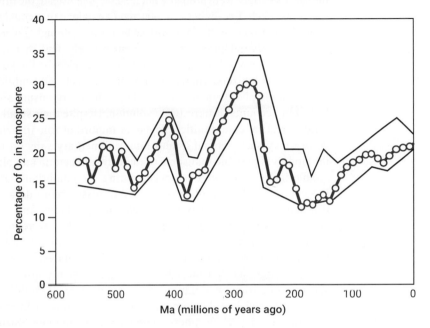

and spread across the world. The high photosynthetic productivity of these new land plants caused profound changes in global biogeochemical cycles such as further increases in atmospheric O_2 levels and the amount of carbon sequestration, plus significant climatic effects (see Fig 6.3). For example, the high levels of carbon sequestration removed so much of the CO_2 from the atmosphere that it probably caused a period of global cooling and the extinction of many species around the world. Finally, we will see how these environmental factors, coupled with the emergence of a rich and diverse terrestrial fauna, resulted in new evolutionary adaptations that led from small and relatively simple bryophytes, such as mosses, to the angiosperm megaflora, especially the forests, that still dominate much of today's global photosynthetic production.

6.3 Early terrestrial colonization: from 1.1 Ga to 600 Ma

There is genomic evidence of the emergence of a rudimentary land biota as early as 1.1 Ga. By 600 Ma some algal protists and cyanobacteria established relatively simple ecosystems located in a few scattered terrestrial regions with regular access to moisture. With the recent discovery of fungal-like fossils dating from about 800 Ma, it has also been proposed that fungi were among the earliest land colonists. The presence of algae and cyanobacteria is also supported by carbon isotope ratio data showing evidence of photosynthetically-driven carbon fixation into biomass on the Earth's surface during the early Proterozoic, possibly as early as 850 Ma. An interesting consequence of these findings is that the first photosynthetic organisms to colonize the Earth's surface were probably not land plants. Instead, the original pioneers were most likely free-living cyanobacteria and algae that could withstand periodic dry episodes on the land surface by forming drought-tolerant spores.

These terrestrial pioneers were comparatively few in number and did not have the ecological and environmental impacts of the subsequent embryophyte colonizers. Therefore, despite the likely 500-million-year reign of non-plant photosynthesizers on land, these organisms are often overlooked in accounts of terrestrial evolution. Despite their small numbers, the early terrestrial communities played an important role in helping to break down rocky surfaces and establishing rudimentary soils on the surface of the land. It is possible that some of these cyanobacteria and algae also lived symbiotically with fungi as lichens during the Cambrian Period or before. However, one of the problems with using fossil record data in this area is that lichens have relatively non-specific morphologies and the presence of lichen-like structures do not necessarily mean they were actually derived from lichens. Indeed, molecular clock data suggests that lichens may not have evolved until very much later. One recent analysis of several hundred genomes suggests that most lichen groups probably arose well after 400 Ma, at a time when vascular plants were already well established.

The early soil-based communities eventually developed into more complex species-rich rhizosphere ecosystems. A complex rhizosphere, with its associated plant-associated symbionts such as fungal mycorhizzae and nitrogen-fixing bacteria, became increasingly important for the later diversification and global spread of the true land plants. Today, approximately 85% of land plants participate in fungal symbioses with various *Mucoromycota* spp., and it is likely that such symbioses were one of the keys to the success of the earliest land plants. By about 500 Ma, increasing atmospheric O_2 concentrations enabled the establishment of a more stable and effective stratospheric ozone layer. This more concentrated ozone layer blocked most of the high-energy radiation that would be otherwise have particularly harmful to the increasingly complex multicellular aerobic organisms that had already diversified in the oceans by this time. In turn, this facilitated the movement of some of these organisms onto the relatively sparsely populated land surface.

By 1.1 Ga, several groups of streptophyte green algae had moved from their original saline oceanic habitats into freshwater habitats that included lakes and rivers. During the >500-million-year period from the appearance of the earliest terrestrial life at 1.1 Ga until unequivocal fossil evidence of true land plants after 500 Ma, it is likely that there was a gradual process of adaptation to semi-aquatic conditions as algae in ponds and lakes were subjected to a combination of wet and dry regimes. These erratic water conditions exerted a selection pressure on freshwater algae. Having originally evolved as completely aquatic organisms these algae now needed to develop drought tolerance and other mechanisms to withstand the rigours of dehydration as their habitats sporadically dried out. One factor that favoured the evolution of land plants during this period was the gradual drying up of the extensive network of shallow lakes, seas, and rivers that had formerly covered most low-lying areas of the principal continental land masses. Only those algae that were able to adapt to a partial or complete drying out of such habitats would survive under these conditions.

The green algal group that most resembles true land plants (embryophytes) is the streptophytes, which includes the charophytes and zygnematophytes. These algae were probably already present in isolated semi-aquatic terrestrial niches by about 1.0 Ga. However, they were still relatively soft aquatic organisms and contained very few structures that could have been preserved in the fossil record. This means that there is no physical evidence of their presence on the pre-Cambrian land surface. More durable external protective structures gradually appeared as their adaptation to drier conditions progressed. One of the first such structures was desiccation-resistant outer coverings on their spores. This would have allowed algae living in areas of ephemeral water presence, such as ponds or shallow lakes, to release large numbers of drought-tolerant spores as their habitat dried out. The spores would then lie dormant until they were able to germinate once wetter conditions returned.

6.4 Transition from aquatic algae to land plants: from 600 to 450 Ma

In Chapter 5, we saw that one of the major groups of green algae, namely the chlorophytes, has maintained a completely aquatic lifestyle where they are still important contributors to global photosynthesis. However, the second major green algal group, the streptophytes, evolved into a much more diverse range of species, some of which made the momentous transition from the water to the land, as shown in Fig 6.4. The streptophytes probably date from before 1.0 Ga. But even at this early stage of their evolution, when they were still all unicellular, some species had already evolved the auxin and abscisic acid signalling pathways that were to play a crucial role in the eventual evolution of multicellular land plants. The earliest land plants were probably derived from branched filamentous algae living in the steadily shrinking lagoons, lakes, and ponds as sea levels fell about 500 Ma. Because these bodies of water were subject to periodic drying, a strong selection pressure would have been imposed on their resident algae that would have favoured variants able to tolerate a lack of water and an increased exposure to the air.

Genomic data have revealed that streptophyte algae contain many gene families that were hitherto regarded as specific to plants. Examples include genes involved in complex cell wall biosynthesis. Complex cell walls are important structural features that enable land plants to develop more rigid bodies able to withstand the force of gravity in a way that is not necessary for aquatic organisms. This suggests that, at the genomic and metabolic levels, many features seen in today's embryophytes had already emerged in their algal ancestors. Other genomic studies show that several groups

Fig. 6.4 Overview of evolution of the streptophytes from aquatic green algae into fully land-adapted plants.

At some time prior to 1 Ga, the green algae split into the chlorophytes and streptophytes. While the chlorophytes remained fully aquatic, some streptophytes went on to colonize the land, probably via aquatic freshwater habitats such as ponds, lakes and rivers. Extant streptophyte algae include the Klebsormidiophyceae, Chlorokybophyceae, and Mesostigmatophyceae (KCM), Zygnematophyceae, Coleochaetophyceae, and Charophyceae (ZCC) groups. Modern streptophyte algae are found in freshwater and terrestrial habitats, for example, in wet soil or on rock surfaces, in the sediment of lakes and streams (Charophyceae), and on the water surface as algal mats (Zygnematophyceae). The land plants are most highly related to the modern ZCC algae that includes charophytes and zygnematophytes. It is likely that one of the important adaptations of these algae was to form beneficial interactions with microbiota in the substrate, including the ancestors of mycorrhizal fungi. The algal ancestor of embryophytes also developed physiological adaptations that allowed it to cope with terrestrial stresses, such as high ultraviolet and photosynthetic irradiance, drought/desiccation, gravity, and rapid temperature changes.

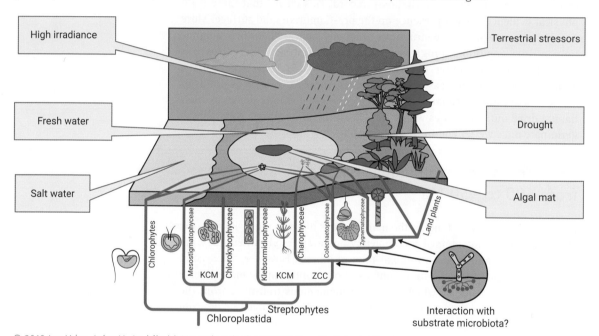

of streptophyte algae have probably acquired genes related to terrestrial-related stresses via horizontal gene transfer from soil bacteria. Examples include signalling-related activities and microbial terpene-synthase-like genes. This is evidence of intimate associations, possibly involving symbiosis and/or syntrophies, between early terrestrial algae and other organisms in the developing rhizosphere that would be important for the later emergence of true land plants. Some of the important genetic and physiological changes that occurred during early plant evolution are shown in Fig 6.5.

In view of this evidence it is probably inaccurate to regard the emergence of land plants in terms of a straightforward movement from their existing aquatic habitats resulting in the colonization of new terrestrial habitats. Instead, the process can be better considered as a series of gradual adaptations by some streptophyte alga that found themselves in aquatic habitats, such as terrestrial lakes and rivers. Unlike the more stable marine

Fig. 6.5 Some key genomic and physiological changes during early land plant evolution.

Some aspects of land plant physiology, such as photorespiration and hormones such as auxins, were already present early in algal evolution. However, most of these features appeared only in the strepto-phyte group where branching and plasmodesmata are found. The appearance of a cuticle and embryo-genesis occurred in the common ancestor of bryophytes and vascular plants and these features are not found in any algae. The hormone gibberellic acid only appeared in vascular plants while brassinoster-oids are only found in seed plants. The appearance of these and other features is shown in green. In a few cases, some of these features were secondarily lost, as shown in red.

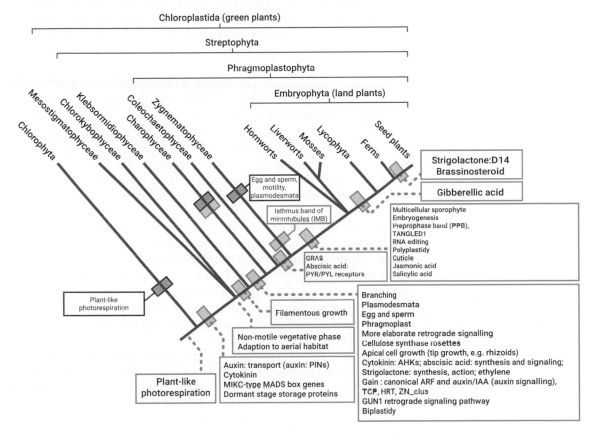

habitats, terrestrial aquatic habitats are subject to novel stresses such as higher radiation levels (including visible light and heat), more frequent pe-riods of dehydration, and more erratic and generally poorer access to many nutrients. These changeable conditions would have favoured the survival of those algae able to develop adaptations to such stresses. Some of the most important adaptations included the ability to withstand drying, eg via fully desiccation-proof spore coats and the presence of waterproof cuticles around its body. Others included the development of new cell wall-based support structures to enable the exposed aerial portions of the body to re-main upright in the absence of a supporting aquatic milieu.

Although some of these algae successfully developed adaptations useful for existence on dry land, this was not a one-way process and, in several cases, some of the land-adapted algae later returned to an aquatic existence. This has been revealed by recent genomic analysis of species from two extant groups of early diverging streptophyte algae, *Mesostigma viride* and *Chlorokybus atmophyticus*. This shows that these, and probably other, streptophytes acquired genes related to terrestrialization, only to lose them later as they secondarily reverted to aquatic lifestyles. In most cases, however, the land-adapted algae probably remained as terrestrial species. It illustrates that there is no prescribed directionality about the movement of photosynthesis from water to land. As with other evolutionary processes, these were contingent events leading to selection of the best adapted variants in a particular set of environmental conditions.

> It is widely accepted that all extant land plants are part of a monophyletic group descended from a single streptophyte algal ancestor that probably emerged onto the supercontinent of Gondwana around or shortly after 550 Ma.

The streptophyte group that eventually made a successful transition to dry land was similar to the present day zygnematophyte and charophyte groups. Phylogenetic studies show that although extant zygnematophytes are more closely related to the land plants, they have undergone a structural simplification from what was probably a more complex plant-like ancestor. In contrast, most extant charophytes have features that are clearly a mixture of algal- and plant-like. For example, Fig 6.6 shows the life cycle and tissue structure of the modern charophyte, *Chara braunii*. In contrast with other green algae, which have relatively simple body plans, both extant and fossil charophytes have complex cellular and tissue features similar to those found in land plants.

In the extant charophytes, these features include plant-like microtubules, apical meristems, plasmodesmata, intricate branching thalli, sporopollenin-enclosed spores, tissues made up of three-dimensional arrays of related cells produced by asymmetric cell divisions, and a placenta that nourishes a retained diploid generation. Unlike other green algae, which have a single plastid that often takes up the majority of the cellular volume, charophyte cells normally contain many smaller plastids. This feature allowed for the expansion of vacuolar organelles, which went on to become sites of storage and waste disposal. In other algal groups, waste disposal involved simply expelling unwanted products into the aquatic medium, which is not an option for a multicellular land-based organism. The smaller size of plastids also resulted in an increased surface area to volume ratio that enabled higher rates of gas and metabolite exchange with the cytosol that facilitated more rapid overall growth rates.

The development of highly regulated intercellular pores, or plasmodesmata, enabled adjacent cells to communicate more efficiently with each other and to exchange nutrients in a highly controlled manner. By linking a whole chain of such cells together via their plasmodesmata, it became possible to establish a mechanism of vectorial transport that eventually led to the phloem system. The phloem is a hugely important feature that underlies

Fig. 6.6 *Chara braunii*, **an aquatic alga with some plant-like features.**

The adult stage of *C. braunii* is a haploid, plant-like structure made up of an aerial structure, called a thallus that has an erect upward-growing stem, and numerous branchlets consisting of lines of single cells connected by plasmodesmata. The thallus and branchlets are both photosynthetic, unlike the rhizoid which is an underground root-like organ made up of several filamentous lines of slender, elongated non-green cells. The diploid phase of the life cycle occurs on the haploid thallus. Here, a male antheridium produces sperm cells that fertilize an egg cell within the female oogonium to produce a short-lived diploid zygote, which is able to undergo dormancy, especially in adverse environmental conditions. Following germination, the zygote undergoes meiosis to produce a new generation of haploid plantlets that grow into adults.

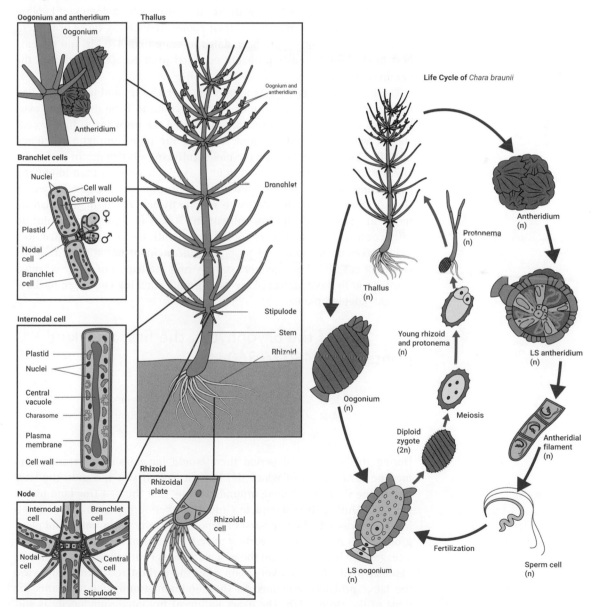

the later success of vascular plants, especially larger species such as trees. It enables the rapid transport over long distances of photosynthetic assimilates from their sites of production in leaves to non-photosynthetic tissues such as roots. Structural features such as apical meristems and the capacity for branching and cell division in different planes also made it possible for the early proto-plants to form a wide variety of body plans. Most important, however, were those adaptations that enabled the plants to survive the sometimes extended periods when their aqueous milieux dried out. As they started to colonize the land, green algae with these useful characteristics developed into small liverwort-like organisms with a dominant haploid generation. Fossilized remains of their spores, resembling those of modern mosses and liverworts, have been dated to more than 470 Ma and the earliest land plant colonists probably emerged prior to 500 Ma, during the Cambrian Period.

The branch of streptophyte algae that gave rise to land plants also had the beginnings of thylakoid membrane stacking to form grana. The evolution of granal stacking probably facilitated the presence of highly red-shifted variants of PSI, without reducing the photosynthetic efficiency of PSII. This feature would have had significant advantages in terrestrial environments because more red light is available on land than in aquatic environments due to the preferential absorption of red light by water. Granal stacking and the ability to efficiently use red light was also an important adaptation during subsequent stages of land plant evolution after communities, such as canopied woodlands, became established. Red light makes up a significant proportion of incident radiation within the shaded areas under a vegetative canopy and the leaves of shade-adapted plants tend to have increased levels of granal stacking compared to those of light-adapted plants.

6.5   Age of the bryophytes, the first true land plants: from 450 to 360 Ma

It probably took several hundred million years for land-based algae to evolve into what are regarded as the first true land plants, the bryophytes.

During this transitional period there would have been numerous intermediate forms, most of which did not succeed. It was not until the later part of the Cambrian Period, around 500 Ma, that the first true land plants emerged. During the subsequent Ordovician Period, between 500 and 440 Ma, there was a distinctive burst of genomic novelty in the early land plant groups. This resulted in the earliest ancestral forms of the two major extant lineages, the bryophytes and the vascular plants (tracheophytes). Although vascular plants emerged very soon after the earliest recorded bryophytes, the latter probably remained as the dominant group of land plants for most of the Ordovician. The exact nature of bryophyte phylogeny is still disputed, but recent sequence data support their monophyly, whereby the three groups, mosses, liverworts, and hornworts, all arose from a single ancestor, while the vascular plants form a separate monophyletic group. As shown in Fig 6.1, it has been proposed that the bryophytes and vascular plants are both descended from a common stomata-containing ancestor called a stomatophyte, although some other phylogenies resolve the liverworts into a separate group from mosses and hornworts. For simplicity, all three groups of non-vascular plants are called bryophytes here.

Despite their modest size and low productivity compared to most vascular plants, it is believed that the bryophytes had a profound influence of the Earth's climate by helping to trigger a widespread glaciation event at the end of the Ordovician. This was at least partially caused by increased weathering of rocks by bryophytes, resulting in a sharp decline in atmospheric CO_2 concentrations. The removal of this powerful greenhouse gas led to a runaway global cooling event that covered most of the land surface in glacial ice. While they are much less abundant today, bryophytes are still an important component of the so-called cryptogamic community of photosynthetic organisms. This group, which also includes cyanobacteria, algae, and lichens, is currently responsible for about 7% of net primary productivity by terrestrial ecosystems. Despite their relative dependence on damp habitats, the Ordovician bryophytes might have occupied as much as half of the terrestrial land area compared to today's vegetation. However, they were gradually displaced from much of this area by the better adapted vascular plants during the Devonian Period after 410 Ma.

The earliest unequivocal fossil remains of intact land plant bodies date from about 430 Ma, but these fossils are not from bryophytes. Instead, they are from the genus *Cooksonia*, which is an early form of vascular plant. It is likely that the relatively small and soft-bodied bryophytes were not sufficiently durable to leave fossil remains, although there are putative plant spores dating from 480 to 465 Ma that could be from early bryophytes. Such spores have been found in several locations around the world, suggesting that early plants might have already been widespread around the late Cambrian and early Ordovician Periods. This is supported by molecular clock data suggesting that the embryophyte clade probably diverged from its algal ancestors between 515 and 470 Ma. The timing of the divergence of the early embryophytes that occurred from the late Cambrian onwards is similar to that of Glomeromycotina, which are the main fungal group with which they formed mycorrhizal symbioses. Because the bryophytes (and their algal predecessors) lacked roots, their fungal partners occupied regions in the main plant body, or thallus, from which rhizoids developed alongside fungal hyphae. This fungal symbiosis was an important part of the success of all subsequent groups of land plants.

One of the most important challenges for the early bryophytes was exposure to full sunlight, with the attendant dangers to their photosynthetic apparatus, especially from photobleaching due to overstimulation of the photosystems. Genomic and biochemical analyses of the moss, *Physcomitrella patens*, has revealed that additional antenna polypeptides are present compared to streptophyte algae. These pigment-protein complexes have a photo-protective role that enables mosses to continue to photosynthesize efficiently in full sunlight. Similar mechanisms are found in all other land plants including seed plants, which suggests that these important adaptations to increased levels of solar radiation were already present in the common ancestor of bryophytes and vascular plants. A structural feature that was important for later land plant evolution was the presence on their outer surface of numerous pores called stomata. The stomata can either be open or closed and their main role is to facilitate gas exchange whereby CO_2 is taken up by the plant while O_2 is released into the atmosphere. In bryophytes, the stomata are relatively peripheral features that are absent from the main plant body and only found on sporangia. They are also expendable and have been lost on at least 63 independent occasions.

Like their immediate algal forebears, the earliest land plants probably lacked all but the most rudimentary support structures. However, competition for space and light would have led to the development of more effective support structures for short stems, based on thickened cell walls. These plants diversified into the multiple bryophyte lineages that include all the extant mosses, liverworts, and hornworts as well as many extinct forms. However, for several reasons, the bryophytes were restricted in their ability to occupy more than a relatively narrow range of ecological niches. One of the key constraints that affected bryophytes was an inability to increase their body size beyond a limited extent so that all extant species are relatively small. There are several reasons for this. Their dominant haploid generation means that bryophytes are unable to increase in size or complexity as rapidly as diploid organisms. Their lack of roots or a vascular system hampers efficient nutrient uptake and assimilate partitioning beyond a relatively small maximum size. Their relatively unstrengthened bodies also preclude aerial growth beyond a few centimetres. And finally, their lack of a continuous impermeable cuticle restricts most bryophytes to moist, shady habitats close to water, which they also require for reproduction.

In order to achieve the wider colonization of the land surface apart from wetlands, many modifications to the bryophyte body plan were required as is seen in the vascular plants.

6.6   Emergence of early vascular plants and the first forests: from 360 to 250 Ma

The vascular plants, or tracheophytes, are a monophyletic group that includes the ferns, gymnosperms, and flowering plants. Ferns and other pteridophytes, such as lycopods, are vascular plants that do not produce seeds, while gymnosperms produce unenclosed seeds, and angiosperms produce fully enclosed seeds. As described in Case study 6.1, vascular plants developed a diverse range of characteristics that enabled them to colonize most terrestrial environments, from the hot tropics to cool boreal regions, and from arid deserts to zones of almost constant rainfall. Some vascular plants even returned to an aquatic existence in various freshwater habitats. Fossil remains of some early vascular plants have been dated from 428 Ma with molecular clock data suggesting an even earlier origin at about 450 Ma,

Case study 6.1
How vascular plants colonized the world

The first land plants were the ancestors of today's bryophytes and were relatively small, soft bodied organisms that were still dependant on water and therefore restricted to habitats such as the margins of rivers and lakes. The real colonization of the majority of the land surface was only achieved by the vascular plants, which shared a common ancestor with bryophytes but developed numerous addition features needed for their terrestrial success.

One of the most notable changes was development of a dominant diploid, sporophyte generation where the dual set of genes enabled a larger number of alleles to persist in a population or species. This meant that the adult plant was a diploid capable of faster growth and larger body size compared to the haploid adult stage of bryophytes. The diploid sporophytes then developed well-defined apical meristems that allowed for the production of discrete organs within the larger plant body. The capacity for shoot meristem proliferation enabled branching of the sporophyte body, as seen in modern vascular plants and some fossil groups known as pre-tracheophytes. This growth habit gave vascular plants far better access both to subterranean resources (nutrients and water) and aerial resources (light and CO_2).

Branching enabled vascular plants to increase their body size, productivity, and reproductive potential as well as the capacity to continue growth, even if some stem cells were damaged or lost (eg by herbivory). Multiple growth points also permitted the specialization of branch systems to form structures such as leaves, cones, and flowers. Branching also facilitated the transition to a dominant sporophyte generation and led to the substantial reduction in size and complexity of the haploid gametophytic body in vascular plants. Many of these plants have augmented their genetic diversity even further by repeated polyploidization to create tetraploids, hexaploids, and even octoploids. Modern breeders routinely create polyploid inter-specific hybrids in order to introduce new genetic diversity into crops.

Partly thanks to their increased genetic diversity, vascular plants have developed much more complex metabolic networks than either bryophytes or algae. This has enabled them to synthesize a greatly extended range of compounds such as phenolics and flavonoids. Some of these compounds allowed vascular plants to develop key structures, such as water-impermeable cuticles, tough lignified cells, and more extensive differentiation into a wide variety of tissue types. These adaptations allowed them to diversify into a wide range of body plans, from small herbaceous grasses and bushes to huge trees able to live for many hundreds of years. Their genetic diversity also enabled vascular plants to develop many novel regulatory mechanisms involving processes such as responses to oxidative stress and heat shock that enabled them to colonize almost any terrestrial habitat, from arid semi-deserts to arctic tundra.

Fig. 6.7 Evolution of the early vascular plants (tracheophytes).

Following the age of the bryophytes from about 500 to 450 Ma, the vascular plants became the dominant vegetation, with groups such as club mosses, horsetails and ferns dominating the Tracheophytic landscape that lasted from about 450 to 390 Ma. This was followed by larger arboreal pteridophytes, such as lycopods, that dominated the Lignophytic landscapes of the mid-late Devonian, from about 390 to 358 Ma. As described in the main text but not shown on this diagram, the pteridophytes continued as the dominant land flora until the rise of the gymnosperms at about 250 Ma.

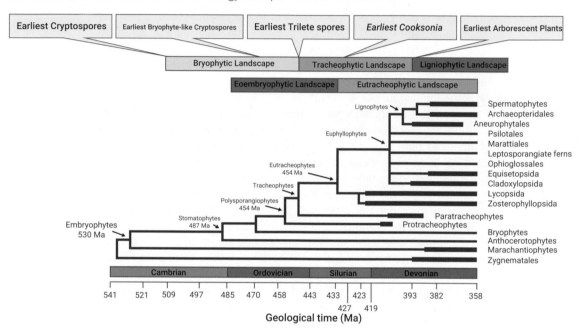

during the late Ordovician. A summary of the main groups of early vascular plants and their relationship with non-vascular land plant groups is shown in Fig 6.7.

> The likely monophyly of bryophytes discussed above means that vascular plants are not derived from a bryophyte ancestor but are a sister group descended from a common terrestrial ancestor.

This has implications for the evolution of both groups since their common ancestor could have had the capacity for free-living haploid gametophytes and diploid sporophytes. In this scenario, within the bryophyte lineage, their diploid sporophytes lost the capacity for free living and became dependant on the haploid gametophyte, which was the main adult stage on their life cycle. In contrast, the haploid gametophytes of vascular plants lost the capacity for free living and became dependant on the diploid sporophyte. A dominant diploid, sporophyte generation was important for vascular plant evolution because the dual set of genes enabled a larger number of alleles to persist in a population or species. Diploidy also facilitated increased body size and, in many vascular plants, this was taken further as they developed stable polyploid genomes that facilitated the evolution of additional forms of diversity, particularly at the level of phytochemistry. Another implication of the common ancestry of vascular plants and bryophytes is that this ancestor might have already developed some form of vascular tissue. This would explain the otherwise puzzling presence in the stems of some moss species of xylem- and phloem-like conducting tissues termed respectively hydroids and leptoids.

Compared to the later vascular plants, the earliest pteridophytes had relatively basic phloem and xylem networks and reproduced via spores rather than seeds. These plants are exemplified in today's vegetation by the ferns and their allies. Some of the earliest vascular plants in the fossil record were the psilophytes, which had no roots or leaves and consisted mainly of photosynthetic stems that looked rather similar to the still-extant whisk fern, *Psilotum nudum*. Thanks to a dominant diploid generation and the beginnings of stronger support structures containing lignin, ferns can attain much larger body sizes and are generally more robust than bryophytes. The earliest forms of lignin date from about 385 Ma. Although these were not as robust as the more complex hardwood-type lignins found in some seed plants, they were sufficiently strong to enable some ferns to produce tall tree-like structures with strong lignified stems or trunks. Extant tree ferns, such as *Cyathea brownii,* can achieve a height of over 20 metres while some Devonian lycopsids grew over 35 metres tall. Unlike bryophytes, all pteridophytes contain stomata, although there are indications that these may not function as efficiently as the stomata in flowering plants. As shown in Fig 6.7, pteridophytes were the dominant flora for almost 150 million years.

By 400 Ma, several groups of seedless vascular plants such as the club mosses, horsetails, and true ferns began to occupy new niches away from the shady damp habitats favoured by most of the bryophytes. The period between 400 and 360 Ma marked the beginning of an extensive colonization of much of the global land mass. During this time the vascular plants partially or completely displaced bryophytes from many of their habitats and were also able to move into new drier regions from which land plants had hitherto been largely absent. Because large browsing herbivorous animals had yet to evolve, vast forests of tree lycopods, tree ferns, and horsetails were able to grow relatively uninterrupted. This resulted in a greatly increased productivity of land plants and the net release of huge amounts of O_2 into the atmosphere.

The period from 390 to 370 Ma was marked by the evolution of increasingly large and more highly lignified trees with deep root systems. Examples include advanced ferns called pro-gymnosperms, such as *Archaeopteris*, which combined characteristics of ferns and true gymnosperms. These plants grew into gymnosperm-like trees up to 24 metres tall with more sophisticated root systems similar to those of seed plants, although they reproduced in the same way as ferns, via spores rather than seeds. Another dominant group of this period was the lycopsids, or club mosses, many of which developed into large arborescent (tree-like) plants. These formed as much as two-thirds of the tropical wetland forests and contributed a

Fig. 6.8 The first trees and forests in a landscape reconstruction of late Devonian Era, about 360 Ma.

This imaginative reconstruction shows a landscape dominated by arboreal pteridophytes. It depicts an alluvial plain in a small river delta with stands of tall *Pseudosporochnus*, up to 4 m high, with *Protopteridium* in the intermediate shrubby layer and herbaceous *Drepanophycus* and *Protolepidodendron* in the understorey.

Source: Reprinted with permission of the author, Jan Sovák. Licensed under CC BY 4.0.

large proportion of the organic material that was deposited and later fossilized into coal. A reconstruction of a late Devonian wetland landscape from about 360 Ma is shown in Fig 6.8. Although some lycopsids are still extant these are relatively small plants and all of the large arborescent forms were displaced by the gymnosperms during the Triassic Period, between 250 and 200 Ma.

> The spectacular success of vascular plants during the Carboniferous Period of 360 to 300 Ma had two important geochemical consequences, namely an increase in atmospheric O_2 concentrations to over 30%, and the deposition of vast amounts of organic matter that was not broken down by herbivores or detritovores.

Much of this dead matter was compressed into carbon-rich fossilized deposits such as coal and oil. The rise in atmospheric O_2 levels had several causes. Firstly, the absence of large herbivores meant that potentially edible biomass such as leaves was not eaten and immediately respired to CO_2. Secondly, there was very little detritovore activity, for example by soil fungi. These factors meant that once they had died, the highly lignified plant bodies did not get respired to CO_2, but instead remained intact as buried organic matter. Under the prevailing wet and largely anaerobic conditions in the lower rhizosphere, this organic matter would have initially been compressed into peat and eventually further processed into the coal, oil, and natural gas that are the basis of the modern fossil fuel industry.

The reason for the lack of lignin decay during the Carboniferous remains unclear. Lignin decay is primarily a fungal activity and it has frequently been argued that there was a long lag period after the evolution of lignin in plants before fungi developed the capacity to degrade this highly durable substance. However, this has recently been challenged by evidence that lignin-degrading fungi were already present during the Carboniferous. It is also the case that high levels of coal accumulation have continued to occur, even during the past 100 million years, when lignin-degrading fungi were definitely highly abundant. Therefore we might need another explanation of why so much wood was deposited without rotting during the Carboniferous. The higher levels of atmospheric O_2 during this period also affected the terrestrial fauna and contributed to the development of gigantism in several arthropod and reptile groups, with some flying insects, such as dragonflies, having wingspans in excess of 0.7 metres.

The gymnosperms are a monophyletic group that were the first plants to reproduce via seeds rather than spores. Due to their structural and physiological adaptations they were able to displace other land plants, especially pteridophytes, from large regions of the earth and went on to dominate terrestrial vegetation for almost 150 million years.

6.7 Age of the gymnosperms: from 250 to 100 Ma

The success of the gymnosperms was favoured by several climatic factors that were operating around 250 Ma. A dramatic fall in precipitation drastically reduced the habitat suitable for tropical wetland species, including many of the arborescent lycopsids, while gymnosperms were able to flourish in the newly arid and nutrient-poor conditions. At the same time, greatly increased CO_2 levels (from about 300 ppm to 1500 ppm) were beneficial to the relatively slow-growing but long-lived gymnosperms.

The ability to reproduce via seeds instead of spores also has several advantages. Seeds are derived from maternal tissue and protect the embryo that forms the future diploid adult plant. Embryos typically contain nutrient stores, including carbohydrates, nitrogen-rich proteins, and lipids, that nourish it after germination and seedling development. Many seeds have the capacity for dormancy, meaning that they can remain alive but quiescent, often for several years, before germinating once conditions are more favourable. Seeds can also be dispersed widely throughout the environment by various biotic and abiotic mechanisms. For example, a gymnosperm cone or an angiosperm fruit can be harvested by an animal vector and transported over considerable distances. While many of the seeds will be eaten by the animal, some of them will drop to the ground while the fruit is being eaten, while others have sufficiently thick coats to pass intact and viable through the animal's digestive tract.

Seed plants probably originated following a genome duplication event during the late Devonian at about 390 Ma. The ancestral seed plant gave rise to the gymnosperm and angiosperm monophyletic lineages. Gymnosperms initially diversified rather slowly with the first conifers, cycads, and ginkgos not appearing until about 320 Ma. However, the increasingly arid and CO_2-enriched climatic conditions of the late Devonian soon favoured gymnosperms over the existing pteridophytes. By the onset of the Triassic Period at 250 Ma, gymnosperms had almost completely replaced club mosses, horsetails, and whisk ferns and were the dominant vegetation across much of the world. As shown in Fig 6.9, gymnosperms maintained this dominant position for almost 200 million years. They constituted the major terrestrial flora through the subsequent Jurassic Period and much of the Cretaceous until the mass-extinction event of 66 Ma, which also wiped out most of the dinosaurs, and ushered in new conditions that greatly favoured the flowering plants, or angiosperms.

Fig. 6.9 The changing balance of the dominant land plant groups since 450 Ma.

These data only reflect species from verified fossil remains and actual species numbers will be considerably higher. The earliest terrestrial flora was dominated by non-vascular plants, such as liverworts and mosses, but simple vascular plants such as lycopods were also increasingly prominent. Although seed plants had already appeared by 360 Ma, the period from 370-250 Ma was dominated by pteridophytes, especially tree species. Cooler, more arid conditions then favoured the spread of a gymnosperm-dominated flora that lasted until the angiosperm radiation that began about 100 Ma and accelerated after the KT mass extinction event some 66 Ma.

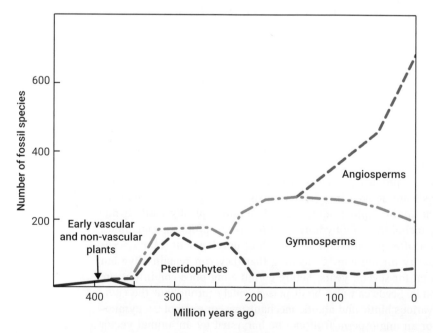

Gymnosperms were the first plants to face substantial threats from herbivorous animals. During the period between 400 and 280 Ma, the vast majority of the early terrestrial animals, most of which were small arthropods, were either predators on other animals or detritovores that fed on dead organic matter. Until the appearance of the large vertebrate herbivores, therefore, most of the energy flow from plants into the animal components of terrestrial ecosystems came via decomposers rather than direct herbivory. After about 260 Ma, the spread of larger and better-adapted herbivores, such as dinosaurs, posed new challenges for terrestrial plants. Vascular plants responded by using their ability to form lignin to increase their height and extent of their protective secondary thickenings, as well as innovations such as thorns. They also used their formidable metabolic versatility to synthesize an increasingly complex range of antifeedants and toxins in order to make their tissues less palatable to prospective herbivores. The first true gymnosperms, including cycads, ginkgos, gnetophytes, and conifers, all date from this period and an artist's reconstruction of a gymnosperm- and fern-dominated landscape is shown in Fig 6.10.

Gymnosperms tend to grow relatively slowly and prefer poorer soils than angiosperms. The slowly decomposing, nutrient-poor litter of gymnosperms tends to maintain these low nutrient levels. Because gymnosperms had arrived first, they created an environment that favoured their slower growth rates and preference for low soil quality. The world from

Fig. 6.10 Gymnosperm- and fern- dominated landscape about 150 Ma.
The age of the gymnosperms lasted from 250 to 66 Ma. This artist's representation shows some of the typical vegetation during the Jurassic Period. At that time the flora was dominated by tall arboreal gymnosperms, such as conifers, cycads, and ginkgos, plus an understory of ferns.

200 to 70 Ma was therefore characterized by a gymnosperm canopy and a fern-dominated understory with a relatively nutrient-poor soil which made up an ecologically stable state. Angiosperms in such a world could only thrive in sites where gymnosperms, and their accompanying nutrient-poor soils, were largely absent or had been displaced, eg following a localized episode such as fire or waterlogging. As we will see in the next section, angiosperms were probably present during much of the extended period of gymnosperm domination, but were unable to displace them until a series of environmental events enabled them to deploy their many novel features, including new forms of photosynthesis, to their full advantage.

The stability of gymnosperm-dominated ecosystems may have precluded the spread of angiosperms beyond a few isolated locations for tens of millions of years. However, once angiosperms in a particular locality reached a critical abundance, there was probably a positive feedback cycle of increasing soil fertility and increasing growth rates. If isolated angiosperm populations were able to link up with neighbouring populations, this might create a 'tipping point' whereby increasing soil fertility tended to exclude gymnosperms and facilitate further extension of habitat colonization by the rapidly growing (and rapidly decaying) angiosperms. Faster growth rates in the more fertile soils would have enabled locally dominant angiosperms to capitalize on their existing reproductive advantages, such as more efficient mechanisms for pollen and seed dispersal.

6.8   Radiation of angiosperms: from 100 Ma to the present

Angiosperms, the flowering plants, first appear definitively in the fossil record at about 140 Ma, although fossilized angiosperm-like pollen has been found dating to 250 Ma. However, genome sequencing data suggest that they might date back as far as 250 or 300 Ma.

Other genomic data from the ancient angiosperm, *Amborella trichopoda*, suggest that, similarly to the origin of seed plants (see section 6.7), a whole-genome duplication event immediately preceded the early radiation of angiosperms, possibly around 160 Ma, although other studies indicate that floral structures were already present on some plants by 180 Ma. Irrespective of its timing, this genome duplication event led to the duplication of several homeotic genes that went on to become the floral organ identity genes responsible for that most characteristic defining features of angiosperms, their flowers. The divergence between the two major angiosperm lineages, the monocots and eudicots, might date from as early as 150 to 140 Ma and by 125 Ma these major groups were already well established.

The major groups of angiosperms and gymnosperms are shown in Fig 6.11. Transcriptome data from 1,000 species of Viridiplantae (mostly angiosperms) show that several genes regulating the development of floral structures were derived from modified versions of much older genes with different functions dating back to the earliest seed plants at 300 Ma.

Fig. 6.11 Angiosperms dominate species diversity among extant land plants.

In terms of species numbers, the contemporary terrestrial flora is dominated by the eudicots and mono-cots, both of which are angiosperms. The other major group of seed plants, the gymnosperms has far fewer extant species, although they still form the dominant flora in many boreal and montane habitats. Among the non-seed plants the bryophytes have maintained a good level of diversity and often form specialised cryptogamic communities with other species. **(a)** Phylogeny of the Viridiplantae, from green algae to land plants. **(b)** Phylogeny of the angiosperms. **(c)** Major sequenced angiosperm species.

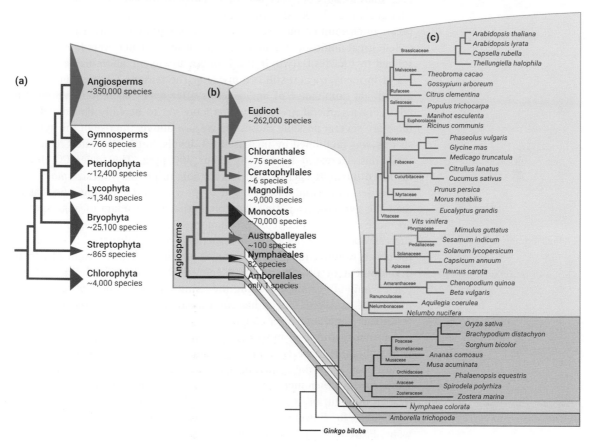

This illustrates what is now recognized as a common theme in evolution, namely that seemingly specific novel functions that arise in a new group of organisms, in this case flowering in angiosperms, do not necessarily require the presence of new gene families. Instead, it is often possible to recruit variants of an existing gene or set of genes to encode the novel function. This process is greatly assisted by duplications of individual genes or even of the whole genome, which is a process that is regularly observed during plant evolution. Once one or more genes have become duplicated, the original copies can continue with their normal functions while the 'extra' copies can potentially mutate without affecting plant viability, thereby creating modified versions that are available for new functions. It is likely that such a process was involved at the very beginning of the evolution of photosynthesis when the original photosystem proteins and the CO_2-fixing enzyme, rubisco, first appeared (see Chapter 3).

By 105 Ma, angiosperms still only accounted for 5–20% of local floras and they were probably limited to a relatively narrow range of habitats, such as aquatic, very dry, or recently disturbed sites. This situation changed around 100 Ma when there was a proliferation of angiosperm taxa, which soon accounted for 80–100% of regional floras. This is the so-called 'great angiosperm radiation', the causes of which were famously described by Charles Darwin in the 19th century as an 'abominable mystery' – and in many ways it still is a mystery. The reasons for the rapid spread and global dominance of angiosperms after over 100 million years as a relatively peripheral group of plants have yet to be fully elucidated, but there are several possible causes. As we saw above, the rise of the gymnosperms was facilitated by increased CO_2 levels, but by the late Cretaceous these has decreased fivefold to about 300 ppm. The lower CO_2 levels favoured the angiosperms, which were able to photosynthesize more rapidly under these conditions, as well as growing and decaying much faster than the gymnosperms. One example is that the stomata of angiosperms are more sensitive to CO_2 changes than those of gymnosperms. Angiosperm success was further helped by new methods of seed dispersal that enabled the plants to move rapidly and colonize new areas. This was aided by their development of prominent, often coloured, seed-bearing fruits that attracted animal vectors, including the rising number of herbivorous mammals that replaced the extinct non-avian dinosaurs as the dominant terrestrial megafauna after 66 Ma.

Another factor contributing to angiosperm success is their coevolution with pollinating insects. Several groups of advanced pollinators such as wasps, bees, butterflies, and moths started to diverge slightly before the angiosperms, but then increased in diversity alongside them. In many habitats insect pollinators can provide a more effective and reliable reproductive mechanism for flowering plants compared to abiotic dispersal via wind or water. Flowering and insect pollination might have originally coevolved in an isolated, possibly insular, location initially with a single specialized partner, such as a wasp. Subsequently, angiosperm floral structures and insect pollinators radiated at rapid rates leading to the enormous diversity that we see today. While there are still examples of specialized insect pollinators and host plants, such as many orchids, most host plants accept a range of pollinators and many insects, such as bees, will pollinate a wide variety of flowers. The greater geographical reach of insect pollination compared to abiotic mechanisms meant that angiosperms were able to exchange genetic information with other conspecifics over long distances. This meant that, despite being sessile, isolated angiosperm populations could interbreed with others located several kilometres away.

Another important factor behind angiosperm radiation might have been their increased ability to form stable polyploids with more complex and versatile genomes. Several of our important staple crops are polyploids, including potatoes, bananas, oats, and the wheat and brassica groups. Polyploidy also facilitated metabolic complexity, especially the ability of the plants to synthesize a wide range of phytochemicals. These compounds include toxins such as alkaloids that can render even soft tissues like leaves unpalatable to herbivores. Such a strategy of chemical defence is much less expensive and more versatile than physical defences such as thorns or thick cuticles. Metabolic resilience promoted by polyploidy might have enabled angiosperms to adapt more readily to the harsh environmental conditions responsible for the KT (Cretaceous–Tertiary) extinction event that was caused by a 10 to 15 km wide asteroid impact in the Yucatán Peninsula. The KT event, which occurred around 66 Ma, resulted in the extinction of about 60% of all plant species, and most animals. However, angiosperm diversity was only slightly reduced by this event and they soon resumed their radiation into the >400,000 species that are known today.

By 66 Ma, plants belonging to modern eudicot families, such as beech, oak, and maple, and the major monocots, including grasses and palms, are clearly discernible in the fossil record. The increasing spread of angiosperms into former gymnosperm-dominated habitats is supported by data from coprolites. These fossilized faecal remains were left by herbivorous dinosaurs that had previously grazed mainly on ferns and gymnosperms. The coprolite data suggest that even prior to their demise around 66 Ma, many herbivorous dinosaurs had adopted a more mixed plant diet that included conifers, cycads, eudicots, and grasses. This indicates that angiosperms were already becoming more abundant before the KT event. As well as colonizing new habitats where gymnosperms were unable to grow, angiosperms displaced the major existing gymnosperm- and fern-dominated floral assemblages across much of the world.

As shown in Fig 6.11, the present-day flora is dominated by angiosperms, as measured by their very high species diversity. While there are only about 1,000 extant gymnosperm species, the angiosperms include >400,000 eudicot species and >80,000 monocot species, including 12,000 graminaceous (grass) species. Eventually, most gymnosperms became restricted to cooler montane, coastal, or boreal locations where the climatic conditions and poorer, often acidic, soils meant that they could still maintain a competitive advantage over angiosperms. Grassland ecosystems, which were dominated by monocots and some shrubby eudicot species, became more common after 20 Ma as several regions, especially the continental interiors, became cooler and more arid. However, true temperate grasslands as seen in contemporary prairie, steppe, and pampa ecosystems probably did not evolve until as recently as 2 Ma.

The angiosperms have developed several significant innovations regarding their photosynthetic efficiency, including C4 and CAM photosynthesis (see Chapters 3 and 7). These can be regarded as physiological mechanisms that enable the plants to concentrate CO_2 for faster growth and/or to survive under conditions of increased aridity. With regard to the light reactions of oxygenic photosynthesis, while numerous minor improvements have been made by angiosperms, they are still essentially using the same basic mechanism with two photosystems that evolved in cyanobacteria well over 3 billion years ago (see Chapters 2 and 3). As we are confronted by a new series of both human and 'naturally' caused biological and environmental changes, we are now attempting to manipulate many useful plant species by substituting Darwinian evolution for bioengineering. Some of the challenges confronting today's land flora and attempts to reengineer photosynthesis for the future are discussed in Chapter 7.

6.9   Loss of photosynthesis in some land plants

Photosynthesis is the most important diagnostic trait for the vast majority of land plants. Despite the evident advantages of this lifestyle, however, a surprisingly large number of plants have either partially or completely lost the ability to photosynthesize.

This evolutionary change is analogous to the loss of photosynthesis found on a much larger scale in many groups of algae (see Chapter 5). The distribution of non-photosynthetic plastids among green algae and land plants is shown in Fig 6.12. In most cases, the plastid genomes are much reduced, but in several cases, including *Polytomella* sp. and *Rafflesia lagascae*, they have been completely lost, although the plastids themselves have been retained, albeit with non-photosynthetic functions. Unlike in the algae, the reversion to a heterotrophic lifestyle in land plants has never resulted in them becoming motile, free-living organisms able to hunt and ingest prey. Instead, all non-photosynthetic plants have remained sessile and are either obligate or facultative parasites of other sessile plants or fungi. The major parasitic mechanism of such plants is the development of modified roots called haustoria that infiltrate into the phloem, xylem, or hyphae of their hosts in order to extract water and nutrients.

Holoparasitic plants typically contain little or no chlorophyll and have highly reduced vestigial plastids. Such plants rely completely on their host and cannot live independently. In contrast, hemiparasitic plants can photosynthesize during some stages of their life cycle, but still require a plant host during other stages. The partial or complete loss of photosynthetic ability has occurred many times in different groups of land plants. It is estimated that this has occurred in as many as 1% of all angiosperms and has evolved independently from free-living ancestors on at least 11 separate occasions. Non-vascular plants have not been as well studied as angiosperms, and there is only a single recorded instance of the loss of photosynthesis in bryophytes, although there are probably other examples yet to be discovered. The parasitic bryophyte is a liverwort, *Aneura mirabilis*, which is able to grow underground in the absence of sunlight where it obtains carbon dioxide from its association with heterotrophic mycorrhiza-forming fungi.

Examples of non-photosynthetic plant parasites are found amongst angiosperms in all of the major terrestrial ecosystems. In some cases, such plants can be serious pests, for example if they parasitize crop plants. In other cases, they might have beneficial ecological functions, such as by reducing the impact of a dominant plant and thereby opening up opportunities for several competitor species. An example of a serious crop pest is

Fig. 6.12 Distribution of non-photosynthetic plastids among green algae and land plants.

The green oval at bottom left represents a photosynthetic primary green plastid bounded by two membranes (two circular lines), which was present in the common ancestor of green algae and plants. Green algae and land plants are divided into two monophyletic lineages: (1) chlorophytes including prasinophyte green algae and chlorophyte green algae (orange branches), and (2) streptophytes including streptophyte green algae and land plants (blue branches). The light blue branches indicates angiosperms including magnoliids, monocots and eudicots. The names of the lineages in which species with non-photosynthetic plastid(s) (white circles, two circular lines) have been found and the species representatives are noted. The asterisks (*) indicate that *Polytomella* sp. and *Rafflesia lagascae* (Malphigiales) have probably completely lost their plastid genomes.

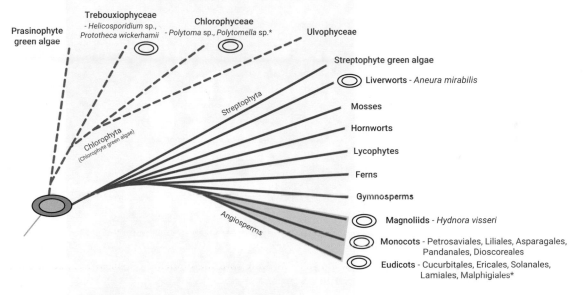

Striga, or witchweed, which is an obligate hemiparasite that is able to infect a wide variety of plant hosts. *Striga* requires a host plant for its germination and initial development, but can then survive on its own as an adult plant. In addition to non-crop hosts, *Striga* infects maize, sorghum, and sugarcane crops and in the USA alone it causes annual damage estimated at well over $20 billion. *Striga* is also a serious pest in the savannah croplands of sub-Saharan Africa, where it causes crop losses worth an annual $13 billion. More seriously, *Striga* reduces the food supply of subsistence farmers who might be driven from the land and face famine as a result.

In addition to parasitizing other plants, there are many examples where plants exploit fungi by acting as mycoheterotrophic parasites. This is especially common in the monocots where numerous examples are found. Sequencing studies have revealed that many mycoheterotrophic eudicot species contain plastids with greatly reduced genomes. Examples include two *Pilostyles* species belonging to the Apodanthaceae family (Cucurbitales), two species from Ericales, the *Cuscuta* genus from the Convolvulaceae family (Solanales), and many species from the Orobanchaceae family

Fig. 6.13 The ghost plant, *Monotropa uniflora*.

This non-photosynthetic plant is a relative of blueberries that parasitises the fungal (mycorrhizal) partners of other plants. It still contains plastids but these organelles have lost all genes related to photosynthesis. Despite this, the plant is still able to produce flowers and attract pollinating insects but it is now a heterotroph that relies on other organisms for sustenance.

Copyright © George Barron (CC BY-NC-ND 4.0)

(Lamiales). One fascinating example of a mycoheterotrophic plant that ultimately parasitizes another plant, but via a fungal intermediary, is the so-called Indian pipe or ghost plant, *Monotropa uniflora*, which is shown in Fig 6.13. The ghost plant contains no chlorophyll or other pigments and is a waxy, ghostly white, hence its name. Its immediate hosts are mycorrhizal fungi, that themselves live in symbiosis with tree hosts such as beech, pine, or oak. Therefore, the Indian pipe is ultimately a parasite of the trees from which its fungal hosts derive their nutrients.

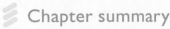

Chapter summary

- The initial phase of land colonization by photosynthetic life occurred from 1.0 Ga to 600 Ma and involved a few pioneering cyanobacteria and streptophyte algae. Between 600 and 450 Ma, streptophyte green algae developed adaptations that led to the first true land plants such as the bryophytes.

- From 450 to 360 Ma, bryophytes dominated land flora although early vascular plants also emerged and from 360 to 250 Ma, a relatively warm, wet period favoured the rise of pteridophytes, including the first trees. High photosynthetic productivity led to atmospheric O_2 levels rising to over 30% and the formation of the Carboniferous coal deposits from the huge amount of plant remains.

- After 250 Ma, a cooler, drier climate favoured the rise of gymnosperms. These seed-bearing plants had improved types of wood that enabled them to form long-lived forest ecosystems. The flowering plants, or angiosperms, may date from 250 Ma or earlier but only achieved their current dominance after 100 Ma. Today, angiosperms make up well over 90% of land plant species.

- About 1% of all flowering plants have partially or completely lost the ability to photosynthesize and have instead become parasites. In some cases, such plants have no leaves, roots, or chlorophyll and contain highly reduced plastids.

Further reading

Bowles AMC, Bechtold U, Paps J (2020) The origin of land plants is rooted in two bursts of genomic novelty, *Current Biol* 30, 1–7. DOI: 10.1016/j.cub.2019.11.090
Genomic data reveal two episodes of genetic novelty crucial to land plant evolution.

De Vries J, Archibald JM (2018) Plant evolution: landmarks on the path to terrestrial life, *New Phytol* 217, 1428–1434. DOI: 10.1111/nph.14975
Detailed but concise review of land plant evolution.

Field KJ, Pressel S (2018) Unity in diversity: structural and functional insights into the ancient partnerships between plants and fungi, *New Phytol* 220, 996–1011. DOI:10.1111/nph.15158
Describes how plant-fungal partnerships have underpinned evolutionary success.

Hadariová L, Vesteg M, Hampl V *et al* (2018) Reductive evolution of chloroplasts in non-photosynthetic plants, algae and protists, *Curr Genetics* 64, 365–387. DOI:10.1007/s00294-017-0761-0
Detailed survey of what happens to plastids when plants become non-photosynthetic.

Kenrick P (2017) How land plant life cycles first evolved, *Science* 358, 1538–1539. DOI: 10.1126/science.aan2923

Brief account of changing plant life cycles from mainly haploid algae to mainly diploid seed plants.

Nelsen MP, DiMichele WA, Peters SE, Boyce CK (2016) Delayed fungal evolution did not cause the Paleozoic peak in coal production, *Proc Natl Acad Sci* 113, 2442–2447. DOI:10.1073/pnas.1517943113

Evidence disputing the hypothesis that coal deposition during the Carboniferous was due to an absence of wood-digesting fungi.

One Thousand Plant Transcriptomes Initiative (2019) One thousand plant transcriptomes and the phylogenomics of green plants, *Nature* 574, 679–685. DOI:10.1038/s41586-019-1693-2

Huge survey of 1,000 plants that informs ideas of their evolution and the importance of repeated genome duplications.

Salamon MA et al (2018) Putative late Ordovician land plants, *New Phytol* 218, 1305–1309. DOI:10.1111/nph.15091

Fossil evidence that vascular plants might have been present as early as 440 Ma.

Soltis PS, Soltis DE (2021) Plant genomes: markers of evolutionary history and drivers of evolutionary change, *Plants, People, Planet* 3, 74–82. DOI: https://doi.org/10.1002/ppp3.10159

Tracing plant evolution by analysing their genomes.

Sousa F, Civan P, Foster PG et al (2020) The chloroplast land plant phylogeny: analyses employing better-fitting tree- and site-heterogeneous composition models, *Front Plant Sci* 11, 1062. DOI: 10.3389/fpls.2020.01062

Recent genetic data suggesting that vascular plants might not have evolved from bryophytes although both had a common ancestor.

Yang Z, Schneider H, Donoghue PC (2018) The timescale of early land plant evolution, *Proc Natl Acad Sci* 115, 2274–2283. DOI:10.1073/pnas.17195881

Molecular clock estimates of early plant evolution showing that most of the major groups were probably already present before 400 Ma.

 Discussion questions

6.1 What were the major factors that enabled one group of aquatic algae to develop into land plants?

6.2 Describe the roles played by climate in the evolution of land plants.

6.3 How and why have some plants lost the ability to photosynthesize and how do they live without it?

7 FUTURE PROSPECTS FOR PHOTOSYNTHESIS AND PLANT EVOLUTION

Learning objectives

- Understanding the huge changes in natural flora caused by human activities.
- Appreciating the global challenges such as food security and climate change that directly involve plants and photosynthesis.
- Using biotechnology and other approaches to improve key aspects of plant performance.
- Future prospects for a sustainable and biodiverse global flora that remains sufficiently productive for human needs.

7.1 Introduction

Photosynthesis is arguably the most important metabolic process on Earth, providing the essential underpinning for the vast majority of life on the planet. During the billions of years since it evolved, the process of photosynthesis, and the organisms in which it functions, have had to adapt to constantly changing ecological, physiological, and geological conditions on a highly dynamic planet. Over the past decade, new evidence from disciplines such as structural biology, genomics, and phylogenetics has revealed plausible mechanisms for the evolution of the key components of photosynthetic electron transport and carbon metabolism. In each of these processes, existing proteins that were likely already present in the last universal common ancestor of life (LUCA) were adapted and modified to serve new functions in photosynthetic metabolism.

While it has yet to be shown conclusively, recent evidence suggests that oxygenic photosynthesis could have been the first photosynthetic mechanism to evolve. If this is the case, it means that non-oxygenic forms of photosynthesis were secondarily derived from oxygenic photosynthesis, rather than being its precursors, as has been widely assumed until recently and as stated in most textbooks. This important debate about the origin of photosynthesis is one of several areas of interest to researchers that might inform a deeper understanding of its role in the evolution of life on Earth. Whatever the timing and nature of its origin, however, there is no doubt that oxygenic photosynthesis is by far the major mechanism of global primary production today. Indeed, non-oxygenic photosynthesis only accounts for 0.03% of global carbon fixation, with oxygenic photosynthesis accounting for the remaining 99.97%.

It is clear that some of the most important protein components of the photosystems had a single origin that traces back to some of the earliest stages in the history of life. Also, several enzymes in the chlorophyll biosynthesis pathway are homologous to nitrogenases, the enzymes responsible for nitrogen fixation, and both groups might have been originally derived from components that were already present in LUCA prior to 4 Ga. And finally, the important CO_2-fixing enzyme, rubisco, is likely to have originated in archaea and/or bacteria as part of an ancient family of enolases already present in LUCA. The evolution of rubisco probably occurred well before the oxygenation of the planet, which means that the carboxylation function of rubisco is most likely to have originally operated in an anaerobic environment. As discussed in Chapter 3, the anaerobic origins of rubisco had important consequences for the efficiency of photosynthetic CO_2 fixation, and these are of great interest in the context of contemporary efforts to engineer more efficient and productive forms of photosynthesis.

The climate of the Earth continues to change in ways that could have significant consequences for life in general, and humans in particular. As discussed in the next section, human activities related to agriculture over the past 11,500 years have already caused a massive impoverishment of terrestrial plant life in terms of both diversity and productivity.

Also, over the past century, there have been huge increases in the combustion of fossil carbon deposits, such as oil, gas, and coal, originally derived from photosynthesis that occurred over 300 million years ago. This has resulted in greatly elevated quantities of atmospheric CO_2 and other greenhouse gases, with CO_2 levels rising from 315 ppm (parts per million) in 1961 to 417 ppm in 2021. The last time these CO_2 levels were reached was over three million years ago at a time when the planet was warmer and wetter than it is today. It is uncertain how increased CO_2 levels will impact future photosynthesis and this is an area of intense research, as discussed in section 7.6 and in Case study 7.1. Methane levels have also doubled over the past two centuries to reach almost 2 ppm, which is significant because its global warming potential is 84 times greater than that of CO_2. The wider climatic consequences of increased greenhouse gas levels are predicted to include higher temperatures in some regions and more erratic

Case study 7.1
How will higher atmospheric CO_2 levels affect crop yields?

Between 1960 and 2021, atmospheric CO_2 concentrations increased from 315 to 417 ppm. This was primarily due to the combustion of fossil fuels for energy and transport. Over the coming decades, continuing release of CO_2 may result in concentrations as high as 550 ppm. The major effect of these CO_2 increases will be on photosynthetic rates in C3 plants which are normally rate-limited by CO_2 availability. It is likely that rates in C3 crops such as wheat and soybeans could increase by as much as 40% if CO_2 levels reach 550 ppm, although increases in grain yield would be somewhat lower. However, this effect would be much reduced in C4 plants because their internal CO_2 levels are already 3 to 6-fold higher than atmospheric concentrations. Therefore, yields of C4 crops such as maize, millet, sorghum, and sugarcane may increase only marginally at higher atmospheric CO_2 concentrations.

Rising atmospheric CO_2 concentrations will affect other physiological processes that affect crop yields. Some pests, such as aphids and weevils, respond positively to elevated CO_2. The quality of the crop may be adversely affected as shown in wheat where higher CO_2 levels resulted in reduced protein content in grains and lower flour quality. As well as encouraging crop growth, higher CO_2 levels will stimulate the growth of other plants, especially weedy C3 species. This could lead to new and more vigorous weeds that threaten crop production. Finally, increased CO_2 levels will tend to reduce water consumption by crops, but any consequent yield gains may be cancelled out by the effects of increased temperatures on evaporation rates.

Given the uncertainty about the extent of future CO_2 increases and the many different ways in which CO_2 can affect plant processes, it is impossible to predict precisely how crop yields will respond on a global basis. In series of papers from the UK Royal Society, it was concluded that overall crop yields would benefit from higher CO_2 levels. However, this yield increase will be offset to some extent by other aspects of climate change, such as reduced rainfall and increased tropospheric ozone levels. Ironically, if as a result of reduced global emissions CO_2 levels are reduced or stabilized well below 550 ppm, crop yields may fall even more rapidly than they would if we maintained slightly higher levels of CO_2.

This analysis of the impact of higher CO_2 levels on crop yields does not give us any clear targets for key traits that could be manipulated to cope with the new conditions. The uncertainty about the impact of CO_2 and other environmental factors reflects our lack of knowledge about the physiological effects of these complex and interactive processes on plant growth and development. The general message is that we need more research into the impact of the environment on a wide range of agronomic traits in all the major crop groups.

rainfall patterns in others. These climatic factors are likely to have mainly negative impacts on food production at a time when global populations are still rising.

Photosynthesis is key to the survival of life on Earth, but the new conditions that we face today mean that it needs to be adapted in order to sustain the necessary levels of primary productivity to support both human life and a diverse flora. The timescale of biological evolution is too slow for such adaptations, but there are good prospects for the use of biotechnological tools to modify some key aspects of photosynthesis. In this chapter, we will look at how our improved knowledge of the major photosynthesis-related processes, such as light harvesting, electron transport, nitrogen fixation, and CO_2 fixation, can potentially be used to engineer more efficient plants. These plants will be required to provide general ecosystem services, such as CO_2 sequestration and O_2 release, plus more specific human-related functions including the supply of food and non-food resources ranging from medicines to building materials and fuel. We will examine several promising biotechnological and synthetic approaches to engineering photosynthesis. Finally, we will look to the future of our global flora, and the photosynthesis that sustains it, as we face many new challenges, most of which have anthropogenic causes.

7.2   Contemporary land flora and future challenges

In terms of biomass and species diversity, today's land natural flora is dominated by the flowering plants, and in particular by eudicots. These plants have evolved during periods of constant environmental upheaval over the past few hundred million years, including several episodes of mass-extinction. In addition to angiosperms, the contemporary land flora includes extant members of several ancient plant groups, albeit in far smaller numbers. Although in many ways the contemporary angiosperm-dominated flora can be regarded as 'more advanced' than previous land plant assemblies, at least in terms of metabolic complexity, this does not mean that today's flora is necessarily more photosynthetically productive than previous assemblies. Net primary productivity is driven by a combination of internal factors in plants, including photosynthetic efficiency. But it is also modulated by a host of external factors such as temperature and water and nutrient availability, as well as by atmospheric O_2 and CO_2 levels.

Due largely to the favourable climatic conditions at the time, it is estimated that global net primary productivity was particularly high during the late Jurassic, at about 155 Ma, when it reached 118 Gt C a^{-1} (giga tonnes of carbon per annum). By the mid Cretaceous, at about 100 Ma, factors such as cooler climatic conditions and lower CO_2 levels resulted in a somewhat reduced productivity of 107 Gt C a^{-1}. Despite the highly efficient and well-adapted angiosperm-dominated terrestrial flora, today's net primary productivity is even lower at 57 Gt C a^{-1}, which is less than half of the value that was reached during the late Jurassic. This much-reduced productivity is largely due to a combination of a contemporary climate that is less favourable to photosynthesis, plus the adverse impact of human activities as we will now discuss.

> One dramatic change that has greatly impacted global flora over the past 11,500 years has been the widespread use of arable and pastoral agriculture by human societies.

This process has greatly accelerated over the past two centuries as populations increased more than eight-fold to almost eight billion people in the 2020s. To feed these extra people, agriculture has expanded into almost every region where crops could be grown, displacing the natural flora and fauna in the process. As shown in Fig 7.1, approximately 38% of the total land area, including many highly photosynthetically productive regions, is now occupied by agriculture. The pastoral land used to feed livestock is mainly open grassland with low species abundance, especially if fertilized. The arable land used for crop cultivation is overwhelmingly occupied by a tiny number of monoculture crops and is even less biodiverse than pastureland.

A further 1.5% of the land is occupied by built-up areas such as cities, meaning that human activities have effectively removed the natural vegetation from nearly 40% of the most productive parts of the land surface. Angiosperm- and gymnosperm-dominated forests, which used to occupy the majority of the global landmass, now only make up about 27% of the total area. Moreover, these remaining productive forests, which are mainly located in the tropics, are becoming increasingly fragmented and affected by human activities that are impacting on their species composition. The 33.7% of land labelled as 'other' in Fig 7.1 is mainly located in relatively unproductive climatic zones, such as cool boreal and montane regions, and arid deserts or semi-deserts. In terms of land plant ecology and species composition, recent human activities have displaced vast areas of eudicot-dominated forest, and have instead introduced a few domesticated crops that are mostly graminaceous monocots.

As shown in Fig 7.2, the progenitors of the graminaceous monocots probably initially diverged from a common ancestor between 65 and 100 Ma. The temperate cereal group, including barley, rye, and wheat, diverged about 10 Ma and their wild ancestors went on to flourish in the rain-fed grasslands of the Near East, where they were first domesticated by humans about 11,500 years ago. The subtropical rices mainly grew originally in South and East Asia and were probably first cultivated as crops about 10,000 years ago. There are fewer widely grown native crops in Africa except sorghum and yams, which have been cultivated since about 5,000 years ago. Finally, the immediate wild progenitor of maize called teosinte, which originated in Mesoamerica, was widely cultivated right across the Americas from about 7,000 years ago.

During the past 500 years all of these crops have undergone a process of globalization, meaning that they are no longer restricted to their various

Fig. 7.1 Global land use and energy dynamics.

(a) This shows the percentages of land covered respectively by arable crops, livestock pasture, forest (wild and cultivated), built-up areas, and others. Much of the cropland in developing countries is still underutilized in terms of its true potential yield. Most pasture land could be converted to arable use by reducing meat and dairy consumption. This might be necessary in order to increase food production for humans as populations grow in the future. **(b)** Solar energy fixation by global photosynthesis in TW (green) and energy consumption of mankind, both metabolically (blue) and technically (brown).

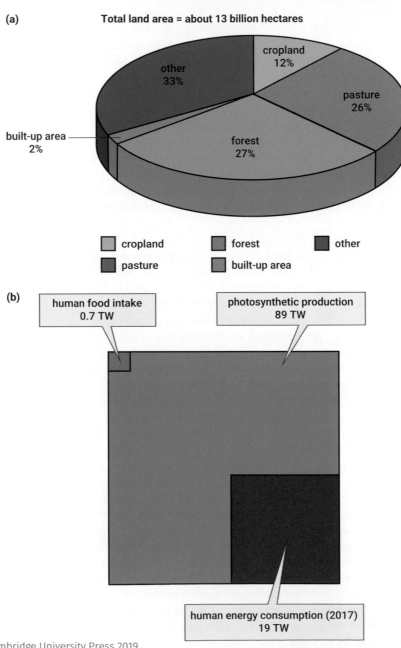

Fig. 7.2 Recent evolution of the major cereal crops.

A tiny number of cereal crops, including rice, barley, rye, wheat, sorghum, and maize, provides more than half of the edible calories consumed by people. Cereals are monocot plants, characterized by long thin leaves, which diverged from the broader leaved dicots about 200 Ma. The first grasses date from about 100 Ma and the most ancient group of cereals are the rices, which appeared about 40 Ma. In the Near East, the major group of cereals were wheat, barley, and rye, which diverged from a common ancestor about 10 Ma. The ancestors of maize and sorghum split about 15 Ma and wild sorghum species became important edible plants in Africa, while teosinte served the same purpose in Mesoamerica until its mutation into the cultivation-friendly form known as maize. The evolution of domesticated forms of these various cereals did not occur until about 11,500 years ago. Although there are still thousands of species of wild cereals, these are now dwarfed in terms of biomass and global distribution by the few cultivated species. Note the logarithmic scale of the time axis. © marks the beginning of domesticated crop cultivation.

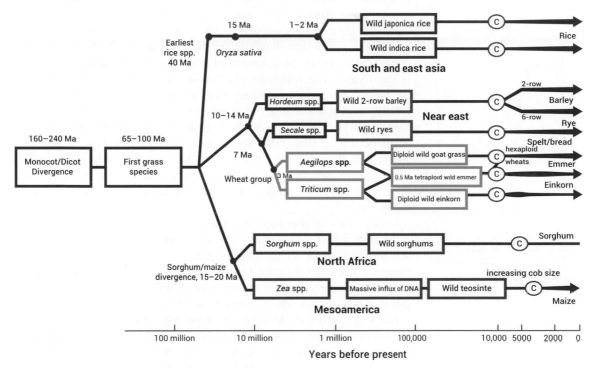

centres of origin, but are cultivated around the world. A tiny number of major cultivated cereals, mostly consisting of wheat, rice, and maize, now accounts for almost half of the entire global crop area and provides about 60% of all food calories consumed by human populations. Much of the remaining cropland is occupied by a few staple eudicot species, including tuber-producing plants such as potatoes, yams, and cassava, or leguminous pulses such as peas, lentils, and beans. While cropland occupies 11.5% of the total global area, pastureland occupies as much as 26.2%. The human-created pastureland systems are used for grazing by a tiny number of domesticated livestock species, such as cattle and sheep, and are generally dominated by just a few grassy monocot species that constitute a near-monoculture. This is especially true where fertilizer is applied to pastureland because the nutrient-rich conditions favour a few rapidly growing grasses at the expense of other plants that would be normally found in an undisturbed meadowland setting.

> Much of the contemporary land flora is made up of a tiny number of human-selected species that are carefully managed to exclude potential ecosystem partners categorized as undesirable.

Examples include other plants that might compete with a particular crop (ie weeds), herbivorous animals that might eat the crop (pests), or microorganisms that might reduce crop yield (diseases). This means that, instead of supporting complex food webs, much of the land has now been engineered for the exclusive purpose of delivering calories for consumption either by humans or their domesticated livestock species. An increasing area of woodland has also been converted to human use as managed tree crops. In many cases, this woodland is a monoculture made up of fast-growing coniferous gymnosperms. These trees produce timber for uses such as building and papermaking, or simply to provide fuel for heating. Such densely-packed woodlands are particularly impoverished in terms of biodiversity as their allelopathic tree needles often exclude most understorey and soil-dwelling species. In a process that is still underway, vast numbers of plant, animal, and microbial species have been excluded from these anthropogenic agro-ecosystems and many have been driven to extinction.

Another habitat that has been drastically affected by human activities is the rhizosphere. Compared with natural ecosystems, the rhizosphere in agricultural systems is highly impoverished in terms of both organic matter content and species diversity. In particular, the reduction in perennial plant species has greatly reduced the amount of mycorrhizae. This is affecting global carbon cycles where plant-mycorrhizal metabolite fluxes generate as much as 7.3 Gt C a^{-1}. Mycorrhizae in the rhizosphere are particularly important for supplying nitrogenous compounds to the majority of plants that do not have symbiotic associations with N_2-fixing bacteria. For most cereal crops, their yields are largely limited by the availability of inorganic nitrogen, and in some cases this has led to the overuse by farmers of nitrogenous fertilizers that are manufactured by highly energy-intensive industrial processes. The runoff of excess fertilizer into watercourses can lead to blooms of algae and cyanobacteria that disturb ecological balances and result in the death of other organisms, such as fish and marine invertebrates.

7.3   Engineering improved photosynthetic light reactions

> Despite its immense evolutionary successes, the overall process of photosynthesis in plants is comparatively inefficient in converting the available sunlight into biomass.

For example, in C3 and C4 plants, the theoretical maximum efficiency for this conversion is only around 4 to 6%, and it is usually well below that in practice. About 75% of the incident solar energy on a plant is lost during light collection because most of it is not captured for photosynthesis and/or is wasted as heat radiation. There are also losses of energy linked to ensuring that light-triggered electron transfer in PSII and PSI occurs only in the forward direction. This loss is an in-built thermodynamic adaptation that protects the system against backwards electron transfer that could lead to harmful ROS production. While some of these energy losses are unavoidable, it is thought that some enhancements or improvements might still be possible. Many pathways and approaches to re-engineer the light-reactions of photosynthesis are available and some of these have already provided promising results. Strategies under investigation include optimizing or extending the light harvesting apparatus, improving or changing the photosystems, and focus on enhancing the ability of plants to protect themselves under variable light environmental conditions that often limit productivity. A summary of some of these strategies is shown in Fig 7.3.

Fig. 7.3 Biotechnological engineering of photosynthetic light reactions.

Proteins from cyanobacteria are shaded blue and plants are green **(a)**, while proteins not associated directly with photosynthesis are orange. Yellow shading indicates non-biological materials. The options shown here include swapping photosynthetic genes from other species **(b)**, **(c)**, using non-photosynthetic genes to create bio-bio hybrids **(d)**, and creating bio-nano hybrids **(e)**, with artificial materials such as photovoltaics. Yet another option is to create completely artificial photosynthetic systems as discussed in Case study 7.2.

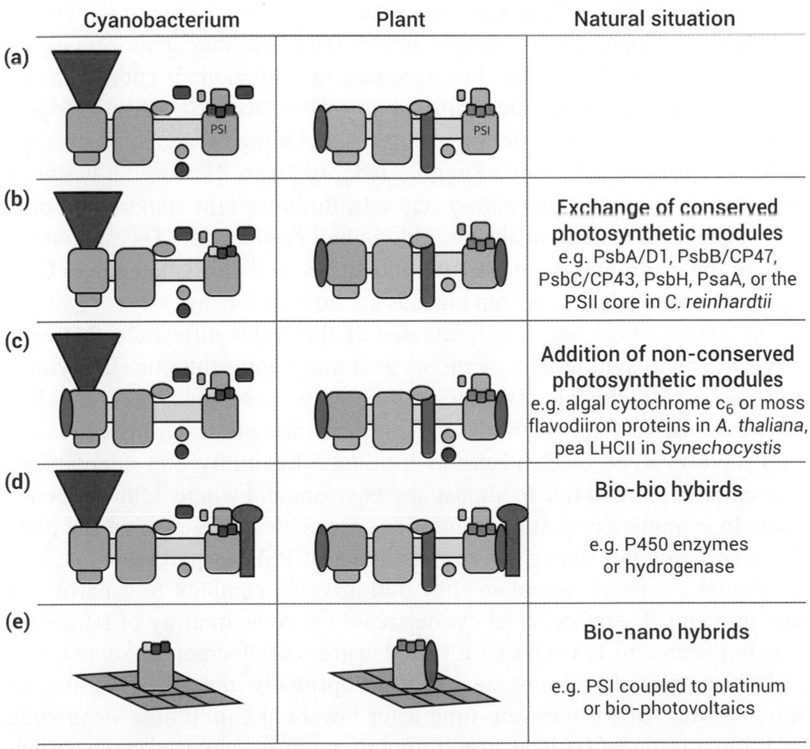

In photosynthetic eukaryotes and cyanobacteria, their photosystems are linked to large light-harvesting complexes that contain hundreds of pigment molecules, such as chlorophylls and carotenoids. It is thought that shading by leaves, even within the foliage of a single plant, can substantially limit photosynthetic capacity. The same is true for algae and cyanobacteria of biotechnological interest grown in bioreactors, because as the cultures become denser, most of the light is rapidly absorbed by cells at the surface of the bioreactor. Decreasing the antenna size could lead to improved and more uniform light penetration. Early attempts to test this hypothesis in liquid cultures of algae gave mixed results, although more recently there have been promising advances in plant systems. For example, in a tobacco variety engineered to contain truncated light-harvesting complexes, biomass production increased by 25%. This was linked to enhanced amounts of PSII, compared to wild-type plants. In another example, rice plants only expressing half the amount of chlorophyll compared to wild-type plants accumulated more rubisco and had higher growth rates.

Oxygenic photosynthesis in most organisms is optimized to use light between 400 and 700 nm, which is often defined as **photosynthetically active radiation** (PAR). It has been proposed that extending this limit into the far-red part of the spectrum (700 to 750 nm) could result in a 19% increase in numbers of usable photons. As we saw in Chapter 2, cyanobacteria have evolved an acclimation response to use far-red light, named FaRLiP. This process involves using a divergent form of D1 to modify the function of PSII so that it can no longer split water to oxygen, but instead oxidizes chlorophyll *a* into the red-shifted form, chlorophyll *f*. Therefore, PSII switches from being a water-splitting enzyme into a chlorophyll *f* synthase. This is achieved by swapping the standard form of D1 for the divergent form called ChlF. It has been shown that the expression of ChlF alone is enough to enable cyanobacteria to produce chlorophyll *f*. The complexity of this adaptation, requiring a large cluster of >20 genes and a highly regulated process, makes it a challenge for it to be engineered into plants. However, it has been demonstrated that just two amino acid substitutions in the standard D1 protein sequence is enough to achieve chlorophyll *f* production. This should facilitate the incorporation of this trait into crops, and efforts to express ChlF genes in marine algae and crop plastids are now underway.

The above adaptation highlights one of the major differences between oxygenic photosynthesis in cyanobacteria and photosynthetic eukaryotes. In general, oxygenic photosynthesis is a highly conserved process that has only evolved very slowly. Both cyanobacteria and photosynthetic eukaryotes have evolved mechanisms to introduce flexibility and adaptability that enables them to live in almost any environment where light can penetrate. In cyanobacteria, such adaptations are achieved by making the photosystem modular. Hence the core subunits of PSII, in particular D1, can be simply swapped around so they optimize the complex to a particular environment. Therefore, most cyanobacteria encode an array of D1 genes, ranging from 2 to 11 copies with varied degrees of divergence. Some forms of D1 fine-tune the photosystem to work optimally under high-light conditions, while other ones fine-tune it for low-light conditions. Meanwhile additional forms of D1 have been found that appear to optimize the system to work under very low oxygen conditions. Differential expression of these and other D1 variants in cyanobacteria has also been associated with other types of stress such as low temperatures or increased UV-B irradiation. It is thought that a divergent form of D1 could switch off PSII function when the cyanobacterium needs to engage in highly oxygen-sensitive processes like nitrogen fixation. So far, only a small number of D1 isoforms have been characterized in detail and many others still remain poorly understood, although they might have useful functions if transferred to plants.

Unfortunately, for as-yet unknown reasons, the impressive photosynthetic versatility shown by cyanobacteria was not inherited by their algal descendants. This might have been due to chance, because the genome of the closest known relatives of the primary plastid, the newly discovered *Gloeomargarita*, only encodes a single D1 gene. This raises the possibility of transferring some of the other forms of D1 from cyanobacteria into eukaryotic plastids with the aim of optimizing photosystem function in a range of environments. Indeed, experiments to replace the native D1 in the model green alga *Chlamydomonas* with high-light or low-light variants from cyanobacteria have shown that mutants expressing D1 variants had more efficient water-splitting cycles that produce more oxygen than the native D1. To the best of our knowledge, no crop plants expressing cyanobacterial variant D1 forms have so far been engineered but these preliminary findings are encouraging. The modularity of PSII is not exclusively associated with D1. For example, *Chroococcidiopsis thermalis* has six versions of D1, three versions of D2, 2 CP43, and 2 CP47 and could assemble up to eight different versions of PSI, as it encodes four PsaB and two PsaA core subunits. What all these variants of PSI do, and how they are expressed, still remains to be discovered.

In natural settings the intensity and quality of light can change rapidly throughout the day. For example, imagine a mixed cloudy and sunny day, or leaves moving in the wind, or ripples in the water working as lenses that suddenly intensify the light for algae. These and other variable conditions can lead to sudden bursts of increased excitation energy that result in the formation of chlorophyll triplet states and the production of ROS. To deal with such risks, photosynthetic organisms have developed mechanisms to prevent or mitigate ROS production. A possible strategy to improve plant resilience against stress and enhance photosynthetic productivity is to enhance these photoprotective mechanisms. For example, a protective mechanism known as non-photochemical quenching (NPQ) can be activated when there is excessive light. NPQ modifies light harvesting so that it switches from a productive photochemical state into a dissipative heat-producing quenching state. Unfortunately, switching back into the productive state can be a prolonged process in plants. However, by overexpressing three enzymes in tobacco plants, the new plants were able to return to the productive photochemical state much faster than wild-type plants. These alterations resulted in up to 20% increased biomass production in the engineered plants.

In addition to these promising approaches to improve photosynthesis in plants by incorporating cyanobacterial components, there are numerous research projects that are focused on developing partially or completely artificial systems. Such systems would be capable of using light energy to generate complex organic products abiotically and would not be limited by a requirement for environmental conditions that have to sustain life. Essentially, they would operate as chemical machines that could be sited anywhere where sufficient sunlight was present. This might sound like a 'science fiction' approach, but recent progress in materials science and our much-improved understanding of the basic photochemical processes of photosynthesis have made it an attractive, if long term, research target. Some of the ways in which artificial photosynthesis might be developed are outlined in Case study 7.2.

Case study 7.2
The quest for artificial photosynthesis

Oxygenic photosynthesis is remarkable in its ability to use solar energy to split water molecules at room temperature. Plants are then able to harness the chemical energy released from water splitting to make complex carbon-based compounds. In contrast, manmade efforts to capture solar energy tend to be relatively inefficient, while the splitting of water can only be achieved at very high temperatures or by using large amounts of electrical energy. This has led to attempts to use hybrid systems such as bio-based pigment protein complexes immobilized on artificial supports to assemble structures capable of carrying out photosynthesis *in vitro*. An example of an approach is the use of $NADP^+$/NADPH coenzymes employing light-generated hybrids and catalysts such as ruthenium for carbohydrate production, ideally from CO_2.

In contrast, totally artificial systems rely on chemistry alone, with no use of bio-derived components, although our greatly improved knowledge of biological water-splitting catalysis can suggest strategies for assembling such devices. Some examples of artificial photosynthesis currently under development include:

- Photoelectrochemical cells using catalysts based on rhodium, cobalt, or ruthenium have been used to mimic the catalytic clusters of enzymes such as the oxygen-evolving complex of PSII and the Ni-Fe or Fe-Fe clusters of hydrogenases.

- Dye-sensitized solar cells using silicon as a photoelectron source and a dye for charge separation and current generation.

- Photocatalytic sheets using cobalt-based catalyst able to generate light-driven conversion of CO_2 and H_2O into formate and O_2 as a potentially scalable technology that combines molecular catalysts and semiconductors.

To date, these and other approaches to artificial photosynthesis have had mixed results. Major outstanding problems relate to the durability of catalysts, difficulty in scaling up the processes, and stability of the artificial cells. Two promising new approaches involve titanium dioxide (TiO_2)- and cobalt oxide (CoO)-based catalysts. In the future, bio- and nanotechnology may enable us to recreate more stable artificial chloroplasts and **biomimetic** approaches will allow more sophisticated versions to be made of existing plant-based systems. The ultimate goal is a new generation of solar powered devices capable of efficient, cheap synthesis of complex carbohydrates, hydrocarbons, and hydrogen for use as fuels, foods, and industrial materials.

7.4   Engineering more efficient CO_2 fixation and concentration

CO_2 fixation

Due to its complex evolutionary history (see Chapter 3), modern rubisco has an exceptionally low catalytic activity and continues to suffer from a highly wasteful side reaction with the second most abundant gas in the atmosphere, oxygen.

This is a less than satisfactory situation for one of the most important enzymes in the photosynthetic process and it seems that rubisco is now trapped in an evolutionary *cul de sac* whereby any improvement in catalytic activity necessarily results in higher oxygenase activity, and vice versa. This has led to research aimed at improving rubisco because future improvements in the activity and specificity of this vital enzyme could lead to dramatic increases in photosynthetic performance including higher crop yields.

There is evidence that, if an improved version of rubisco (and its chaperonins) could be engineered, the rest of the RPP pathway would respond by functioning more rapidly with an overall increase in CO_2 fixation and plant productivity. Currently there are two major approaches under investigation. These are: (i) re-engineering the rubisco enzyme and (ii) developing alternative mechanisms of CO_2 fixation that bypass rubisco altogether. The first approach has so far had only limited success. However, as shown in Case study 7.3 researchers are using several approaches including transferring a more efficient version of rubisco from cyanobacteria to crop plants, or using molecular engineering approaches to redesign a completely new form of the enzyme.

In terms of bypassing rubisco, several synthetic biology projects are aimed at creating novel CO_2 fixation pathways. Examples include inserting the malonyl-CoA-oxaloacetate-glyoxylate (MOG) or the crotonyl-CoA/ethylmalonyl-CoA/hydroxybutyryl-CoA (CETCH) cycle into plants. The latter pathway, for instance, is based on an efficient and versatile class of enoyl-CoA carboxylases/reductases. Unlike rubisco, these enzymes are unaffected by O_2 as they use a different CO_2-fixation mechanism. A complete CETCH cycle using 17 different enzymes from nine different organisms,

Case study 7.3
Engineering a more efficient form of rubisco

Rubisco is arguably one of the most important enzymes on Earth. This enzyme catalyses the photosynthetically driven fixation of CO_2 in cyanobacteria, algae, and plants. It is therefore ultimately responsible for the formation of all the complex carbon compounds in these organisms. Unfortunately, however, rubisco also suffers from the following two serious drawbacks that limit its efficiency and reduce the potential productivity of plants:

(i)   In addition to its primary carboxylase activity, rubisco can act as an oxygenase. For this reaction it reacts with molecular oxygen instead of CO_2 resulting in the convert of its substrate, ribulose bisphosphate, into 3-phosphglycerate and 2-phosphoglycolate in an energy-wasteful side reaction.

(ii) The catalytic rate of rubisco is one of the slowest of any enzyme yet measured. It is so sluggish that, in order to keep up with the demands of CO_2 fixation, photosynthetic cells need to produce vast quantities of the enzyme. In a typical mature leaf rubisco constitutes as much as half the soluble protein, which imposes a huge drain on the metabolism of the plant.

Because of its slow catalytic rate and its poor substrate specificity, rubisco has both puzzled and fascinated researchers for many decades. As discussed in the main text, several biotechnological approaches are currently being taken to improve the efficiency of rubisco in order to improve photosynthetic rates and hence, possibly, to increase crop yields. For example, site-directed mutagenesis is being used to modify the active site of rubisco in model organisms such as the cyanobacterium, *Synechococcus*, and the alga, *Chlamydomonas reinhardtii*.

While these and other approaches have had some limited success, some researchers believe that rubisco is already highly adapted to its present highly oxidizing subcellular environment and that genetic engineering is unlikely to produce more than relatively modest improvements in its catalytic efficiency and plant growth. However, there has been some recent progress in bypassing rubisco and replacing it with more efficient enzymes. While this research is at an early stage and might take decades before coming to fruition, it is important in tackling one of the major bottlenecks to improving photosynthetic performance, especially in food crops.

plus three re-engineered enzymes has recently been assembled *in vitro* and was shown to turn over at a rate comparable to the natural RPP cycle.

The native (RPP) and synthetic (CETCH) CO_2 fixation pathways are compared in Fig 7.4. Biochemical studies indicate that a synthetic CETCH cycle

Fig. 7.4 Two schemes for engineering carbon metabolism to increase crop yields.

In normal photosynthesis (top), photorespiration competes with the RPP cycle when rubisco acts on oxygen instead of CO_2. Photorespiration requires ATP, and CO_2 is released from a carbon previously fixed through the RPP cycle (left). This inefficiency reduces the yield potential of C3 plants (right). Yield losses can be prevented through bypassing the photorespiratory pathway that promotes carboxylation by rubisco in the chloroplast. The first scheme (middle) employs glycolate dehydrogenase from *Chlamydomonas reinhardtii*, malate synthase from *Cucurbita maxima*, and endogenous malic enzyme and pyruvate dehydrogenase for glycolate oxidation with release of CO_2 inside the chloroplast. Deletion (black cross, left) of the plastidial glycolate/glycerate transporter 1 (PLGG1) reduces the export of glycolate from the chloroplast. This enhances glycolate consumption by the alternative bypass allowing carboxylation of CO_2 through rubisco with a proportionate yield increase (right). The second scheme (bottom) shows the crotonyl-CoA/ethylmalonyl-CoA/hydroxybutyryl-CoA (CETCH) cycle integrated with scheme one. Rubisco and the RPP cycle are bypassed by the CETCH cycle, which is a set of synthetically designed, highly efficient carboxylases that can fix CO_2 many times faster than the RPP cycle.

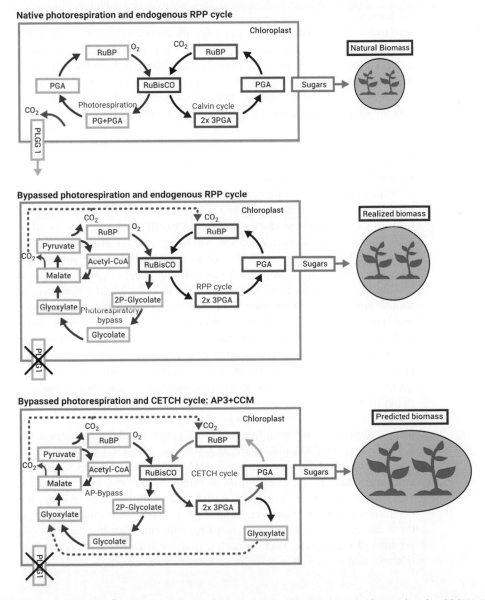

could potentially fix about 80 molecules of CO_2 per second, in contrast to only 2 to 5 molecules with the native RPP cycle. If even some of this impressive gain in CO_2 fixation efficiency could be translated into increased plant productivity, the results could be momentous in terms of food security. Although initial results of such approaches are promising *in vitro*, the transfer of this biotechnology to whole plants will be a significant challenge that remains a future aspiration requiring a lot more research to fulfil.

CO_2 concentration

The main advantage of C4 photosynthesis is its ability to generate high concentrations of CO_2 in the vicinity of the rubisco active site, hence greatly increasing enzyme efficiency (see Chapter 3). C4 plants can generate local CO_2 concentrations of 1,000-2,000 ppm compared to atmospheric levels of 410 ppm. However, this is a highly energy-intensive process and C4 plants only have an advantage over C3 plants in certain conditions, namely high temperatures, low rainfall, and relatively high levels of sunlight. Recent climatic trends are increasing the incidence of such conditions in many crop-growing regions, making C4 photosynthesis an attractive trait to transfer into some C3 species such as wheat, rice, and potatoes. One exception to the advantage of C4 versus C3 photosynthesis is that it is most marked when atmospheric CO_2 concentrations are low and those of O_2 are high. Unfortunately, however, current trends involve increases in atmospheric CO_2 concentrations.

Despite these caveats, there are good reasons to consider biotechnological approaches to transferring the C4 trait to some crops. The most productive C4 crops can produce yields and growth rates as much as 40 to 50% higher than the best C3 crops. There are several projects aimed at converting C3 into C4 crops with rice as a particularly promising target. Rice is a warm weather crop that feeds 3 billion people and it is estimated that its yields will need to increase by 50% in order to feed the growing populations in regions dependant on rice as a food staple. It is possible that transferring the C4 trait into rice could achieve such a yield increase without requiring additional cropland while also improving the efficiency of nitrogen and water use. This particular approach involves the transfer of at least 12 genes in order to set up the biochemical pathway, plus a longer term effort to replicate the specialized vascular bundle sheath anatomy found in most C4 plants.

Like C4 photosynthesis, CAM has evolved independently many times in plants. Although it is found in some pteridophytes and gymnosperms, the vast majority of CAM plants are angiosperms. While C4 photosynthesis is most common in grassy monocots, CAM is mainly found in succulents and cacti where it is an adaptation to relatively hot and arid conditions. This trait is mainly of advantage under arid conditions where crops are not normally grown. However, as aridity increases with climate change, there might be some advantages to using biotechnology to produce CAM versions of some crops, such as barley, that are already relatively drought tolerant, and several such projects are currently underway.

7.5   Engineering nitrogen fixation

There is considerable interest in transferring the nitrogen fixation trait into some of the major crop plants. This is because the overall yields of many plants, especially modern elite crop varieties, tend to be limited by the availability of nitrogenous compounds.

This problem can be addressed by supplementing soil nitrogen with fertilizers, either organic (eg animal manure) or inorganic (eg ammonium nitrate). While these compounds boost crop yields they have several drawbacks that limit their sustainable use in the long term. Organic fertilizers such as animal manure are bulky, expensive to transport, and only available from livestock in quantities that are not sufficient for modern cropping systems. On the other hand, the manufacture of inorganic fertilizers requires vast amounts of energy from fossil fuels, so they are both financially and environmentally expensive. High financial costs mean that many farmers in developing countries cannot afford fertilizers and suffer reduced crop yields. The environmental cost comes from pollution of watercourses caused by fertilizer runoff, plus the release of potent greenhouse gases in the form of various oxides of nitrogen, or NO_x. It is estimated that fertilizers directly cost farmers over \$100 billion/year, while their environmental costs are over \$160 billion/year. Therefore, nitrogen-fixing crops would reduce costs and increase food yields, as well as addressing the environmental problems, thereby making agriculture as a whole much more sustainable.

Bacterial nitrogenases are the only known enzymes that are capable of directly fixing N_2 gas into ammonia. This means that, unlike CO_2 fixation as discussed above, there are no alternative routes to creating N_2-fixing plants apart from existing nitrogenases. There has been much research into the mechanism of nitrogenase catalysis and the feasibility of transferring the enzyme into plants. Two of the most serious challenges for biotechnologists trying to engineer N_2 fixation into crop plants are the complexity of the enzyme system, and its acute sensitivity to inhibition in the presence of oxygen. The best studied nitrogenase is the molybdenum nitrogenase, which consists of an iron protein and a molybdenum/iron protein, although the correct assembly and activity of these proteins require additional chaperone, electron-transfer, and cofactor assembly proteins. The core reaction sequence of molybdenum nitrogenase is shown in Fig 7.5. In nitrogen-fixing bacteria the genes encoding the nitrogenase system are clustered on the *nif* regulon and can range in number from six to more than 50.

Bacterial nitrogenase genes have been expressed in tobacco plants and some activity detected. It has also been shown in a model system that plant-derived electron-transfer proteins can substitute for their bacterial analogues. An important aspect of these studies is the need to minimize the risk of the nitrogenase system becoming irreversibly inhibited by oxygen. To achieve this, one strategy is to localize the entire system inside mitochondria, which have relatively low oxygen levels. Despite some initial

Fig. 7.5 Theoretical scheme for biological nitrogen fixation pathway.

(a) Cycle showing putative intermediates bound to the iron-molybdenum (FeMo) cofactor, abbreviated M, during N_2 reduction by nitrogenase. **(b)** Alternative view showing the consumption of ATP and reduced ferredoxin (Fd_{red}) during this highly energy-requiring process.

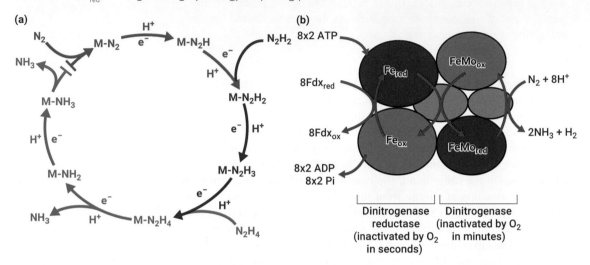

(A) Image provided courtesy of Dennis R Dean © 2006
(B) Copyright © 2013 Elsevier Inc. All rights reserved.

successes, formidable challenges remain before a complete and fully active nitrogenase system can be inserted into model plants such as tobacco. Even if nitrogen fixation is achieved in the coming decades, another issue is that the nitrogenase product, ammonium, can be toxic to plants and requires further nitrification to nitrites and nitrates to render it safe. The extension of this technology to major target crops, such as wheat and rice, will involve yet further challenges. For example, the high energy requirement of nitrogen fixation might cancel out any increase in nitrogen use efficiency, resulting in little or no change in overall crop yield.

In view of the difficulties in transferring the entire nitrogen fixation trait to crops, alternative approaches are also being investigated. One of these involves the reassembly of the naturally occurring nodule-forming symbiosis of legumes and rhizobia into cereals and other crops where it is not currently present (see Fig 7.6). Legumes recruit their bacterial symbionts, such as rhizobia, by releasing flavonoids from their roots. This stimulates rhizobia to produce the signalling molecule *Nod factor* that initiates a series of developmental responses in plant hosts via the common symbiosis, or SYM, pathway. The plants accommodate their symbiotic partners in specialized organs called nodules where the bacteria differentiate into nitrogen-fixing organelle-like structures called bacteroids. These are surrounded by a plant-derived membrane that is reminiscent of the cyanobacterial-derived plastid organelles contained within a host-derived membrane in plants and algae (see Chapter 4).

Interestingly, the rhizobium-legume symbiosis, which is a relatively recent innovation dating from about 60 Ma, has recruited many

Fig. 7.6 Engineering nitrogen fixation into cereal crops.

(a) The nitrogenase enzyme, shown in orange. The holoenzyme is composed of a homodimer of the Fe protein, NifH, and a heterotetramer of the FeMo protein, NifDK. **(b)** Nitrogen fixation occurs in legume crops, which form specialized symbioses with nitrogen-fixing rhizobia bacteria. Legume roots form nodule structures which house the rhizobia and provide them with sugars in exchange for fixed nitrogen. Nodule structures aid nitrogen-fixation by creating an environment rich in photosynthate and low in oxygen through several mechanisms, including production of leghaemoglobin protein, which sequesters oxygen and gives the nodules a pink colour. **(c)** Harnessing nitrogen fixation for cereal crops has been pursued via two major strategies: plant engineering and bacterial inoculants. Plant Engineering involves heterologous expression of nitrogenase genes (nif) in plant cells. It is challenging due to the complex genetics of nitrogenase biosynthesis, as well as the sensitivity of nitrogenase to oxygen. Prospects for engineering nodule-like symbiosis with rhizobia in cereal crops are increasingly promising, as cereal crops contain signalling pathways analogous to those responsible for nodule organogenesis in legumes. Bacterial Inoculants involves using nitrogen-fixing bacteria that naturally associate with the roots of cereal crops. Such inoculants could provide an avenue for providing nitrogen through nitrogen fixation when applied to crops as inoculants.

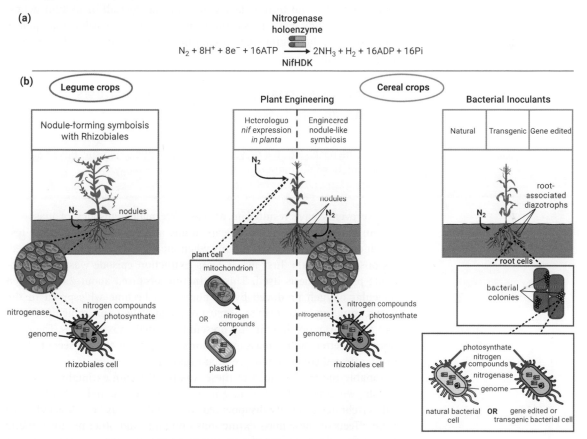

© 2019 Sarah E. Bloch, Min-Hyung Ryu, Bilge Ozaydin, Richard Broglie. Published by Elsevier Ltd.

components from the much older SYM pathway that is responsible for symbiosis with arbuscular mycorrhizal fungi, which dates from before 450 Ma. Unlike the rhizobium-legume symbiosis, fungal symbiosis is widespread in plants including cereals. This means that it might be possible to use modern synthetic biology techniques to tweak the existing fungal-specific SYM pathways in crops so that they can establish symbioses with rhizobia instead. An alternative approach is to engineer an entirely new type of organelle dedicated to nitrogen fixation, sometimes called the nitroplast. Some examples of such organelles exist in nature, including spheroid bodies. These are permanent intracellular endosymbionts of rhopalodiacean diatoms, that are in the process of evolving towards becoming nitrogen-fixing organelles. Spheroid bodies originated relatively recently from a cyanobacterium closely related to the nitrogen-fixing *Cyanothece*. The genome of this endosymbiotic cyanobacterium has lost essential photosynthetic genes for both PSI and PSII, and shows substantial reduction in size and metabolic capacity to such a degree that it is entirely dependent on its host. A similar symbiotic association has evolved independently between another cyanobacterium and a prymnesiophyte alga. In this case, the cyanobacterium is an obligate ectosymbiont that has retained PSI and nitrogenase genes but has lost all of its PSII genes. With the new advances in synthetic biology, it is possible that plants could someday be engineered to contain their own dedicated nitrogen-fixing organelles.

7.6   Future prospects for photosynthesis and the global flora

During the past 3 to 4 billion years, photosynthetic organisms have had to cope with huge fluctuations in the climate and geology.

These events have resulted in at least five major episodes of mass extinction in the last 450 million years alone, and probably many more less well-documented earlier extinctions when most of life was aquatic. Indeed, it is possible that the first major mass extinction episode was due to oxygenic photosynthesis itself. This extinction occurred about 2.4 Ga when molecular oxygen produced by cyanobacteria became widespread in the oceans, resulting in the demise of most of the prevalent anaerobic life forms, apart from survivors that are still present in limited anoxic niches. The more recent and better studied mass extinctions dating from 450 Ma seem to have mostly affected complex multicellular organisms.

Probably the most serious of these was the Permian extinction at about 252 Ma, when 80–90% of marine invertebrate species, and 70% of terrestrial vertebrates suddenly disappeared. To date there has been less evidence of the effects of these mass extinctions on plants and other photosynthetic organisms. Indeed, several recent studies suggest that, at least in the case of the Permian extinction, plants were hardly affected. Although there is relatively little evidence on the impact of other mass extinctions on plants, it is possible that photosynthetic organisms as a whole might be more resilient in responding to these global-scale events. It is useful to bear this in mind when considering the future of photosynthetic life. Also, compared with previous climatic and geological perturbations, the current largely anthropogenic changes are relatively minor in their magnitude. However, their

Fig. 7.7 Changes in global plant productivity, 1982-1999.

The map shows productivity increases from 1982-1999 in green, while decreases are shown in brown. During this period, the climate became warmer, wetter, and sunnier in many parts of the world and CO_2 levels increased everywhere. Satellite observations reveal that these changes increased the overall productivity of land plants by 6%. Primary productivity, which is the net uptake of carbon, increased the most in several tropical regions and northern latitudes. In the tropics, climate change resulted in fewer clouds and more sunlight, while in the North, temperatures increased. Although 25% of the Earth's vegetated surface experienced increased its productivity, 7% experienced a decrease. These results show the complex nature of plant responses to climate change, with decreases due to factors such as less rainfall sometime more than offset by increases due to elevated CO_2 levels.

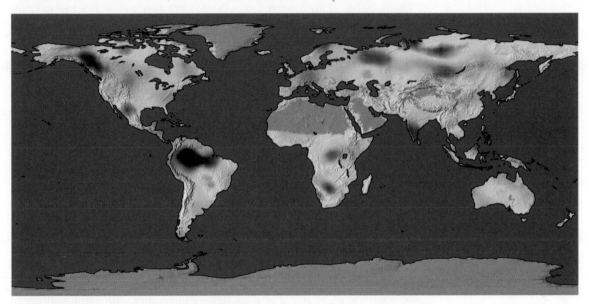

Image by Robert Simmon, based on data provided by the University of Montana Numerical Terradynamic Simulations Group

results could be profound for the human-dominated agro-ecosystems upon which modern life depends (see Fig 7.7).

Due to changes in climate patterns, the environment in many crop-growing regions is likely to alter considerably.

These climatic shifts will probably be more modest compared to some previous changes, such as the ones that created 'snowball earth' on several occasions prior to 600 Ma. However, they could still profoundly affect the balance of the global flora, reducing overall diversity and possibly endangering some of our staple crop species. While it may be possible to mitigate some of the climatic changes in the long term by reducing greenhouse gas emissions, scientists predict that significant changes will still occur during the remainder of the twenty-first century. By 2050, it is predicted that atmospheric CO_2 concentrations could reach 550 ppm and ground-level (**tropospheric**) ozone will reach 60 ppb. Although there will be considerable regional differences, average global temperatures might be about

2°C higher than in 2000 and rainfall will be reduced in some key crop growing regions. Each of these climate-related factors is likely to have some effect on plant productivity, although the impact of these and other climate-related factors on agriculture remains uncertain. This uncertainty makes it difficult to develop specific crop traits that will enable us to cope with climatically related threats, as discussed in the case of elevated CO_2 in Case study 7.1.

Although elevated CO_2 levels could increase crop yields, especially in C3 crops like wheat, many of these gains could be cancelled out by other factors such as drought. However, higher CO_2 levels may give breeders additional opportunities to increase yields under the new conditions. Higher tropospheric ozone levels will decrease crop yield by at least 5%, but it may be possible to select some crop varieties that are less prone to ozone damage. Improved drought tolerance and response to higher temperatures are clearly major traits that should have high priority in breeding programmes. Providing all available tools of plant breeding can be used, including genome editing, overall crop yields could be increased by 50% by 2060. However, this will only be possible by the manipulation, not just of yield itself, but also the indirect effects of climate change and especially the stress-related traits of crops. For humanity, the major challenge for the next century will be to maintain enough crop productivity for feed a population likely to reach 9–10 billion by the late 2000s, although there are recent signs that numbers might gradually decline after that peak.

More broadly, there will be considerable challenges in managing the crisis facing the natural global biota due to anthropogenic factors that are already contributing to what has recently been called the Holocene (or Anthropocene) mass extinction event. There are estimates that as many as one million plant and animal species are threatened with extinction due to anthropogenic habitat loss, pollution, and climate change. In the case of plants, tropical ecosystems are particularly at risk as they contain over two-thirds of all angiosperm, pteridophyte, and liverwort species, while mosses are more uniformly distributed. Within the tropics, the areas of maximum plant diversity are in South America and South East Asia, with lower densities in the African and island tropics. For example, 96% of all tree species are in the tropics, with about 20–25,000 species each in South America and South East Asia, but only 4,500–6,000 in Africa. This might be partially due to the legacy of the higher impact of humans in Africa over the past few millennia. Unfortunately, however, the species-rich tropical regions of South America and South East Asia are now under huge threat due to anthropogenic activities mainly related to agriculture.

This brings us to the dilemma of how can we continue to produce enough food and maintain an adequate standard of living for 9–10 billion people, while simultaneously minimizing the impact of such activities on the rest of global life?

The first priority is to understand the scope of the problem and such work is still in its infancy with fewer than 20,000 plant species out of over a million being formally assessed at the global level. Preliminary estimates indicate that half of all plant species are threatened or near-threatened, with gymnosperms being four times more at risk than legumes. It might be possible to try to achieve targets of 'zero extinction' at local level where there is good research infrastructure, public support, and availability of seed banks and refuges, such as conservation areas. However, such facilities are less likely to be available on the required scale in the critically affected areas, such as the tropics of South America and South East Asia. This means that addressing plant extinctions must be a worldwide endeavour similarly to the response to climate change.

Extinctions are a natural consequence of biological evolution but the extent of current anthropogenic extinctions dwarfs that of normal natural selection. While it will not be possible to save all threatened plant species, we can focus on certain groups, such as plants of economic value (eg wild relatives of crop and medicinal plants), ecological importance (ie key roles in ecosystem functioning), and phylogenetic uniqueness (ie species with no close relatives). In conclusion, there are formidable challenges confronting photosynthetic life on Earth, particularly the terrestrial flora in the tropics. Because most of these are caused by humans, it is our responsibility to seek to conserve such natural resources for future generations. However, it is also true that photosynthetic life has been around for almost as long as the Earth. Indeed, given its resilience it seems likely that, even if humans and rainforests eventually disappear, there will still be robust terrestrial and aquatic floras able to continue the process of photosynthesis for many millions or even billions of years to come.

Chapter summary

- The contemporary land flora is increasingly dominated by human activities, especially agriculture. This has drastically reduced the global area occupied by 'natural' vegetation and replaced it with a small number of domesticated plants used to feed either people or their livestock.

- Several important photosynthetic processes are amenable to improvement via biotechnology. Light-harvesting and electron-transfer mechanisms in crops might be improved by incorporating features of cyanobacterial systems.

- The fixation of CO_2 via rubisco is particularly slow and inefficient leading to efforts either to re-engineer and improved version of rubisco or to bypass it by construction a new CO_2 fixation pathway using synthetic biology.

- As climatic change increases temperatures and aridity in some crop growing regions, the transfer of CO_2 concentration mechanisms, such as C4 or CAM, to major crop species is being attempted.

- Nitrogen availability is a major constraint on crop production and the transfer of the N_2 fixation pathway from bacteria is an important research goal.
- Finally, there are huge challenges facing photosynthetic life on Earth, especially in terrestrial ecosystems. Many plant and animal groups are facing extinction and it is vital that strategies are developed to minimize human impacts.

Further reading

Blankenship RE et al (2011) Comparing photosynthetic and photovoltaic efficiencies and recognizing the potential for improvement. *Science* 332, 805–809. DOI: 10.1126/science.1200165

Comparison of the efficiency photosynthesis with artificial methods of light energy capture and use.

De Souza P et al (2022) Soybean photosynthesis and crop yield are improved by accelerating recovery from photoprotection. Science 377, 851-854. DOI: 10.1126/science.adc9831

Soybean plants were bioengineered to speed up photoprotection via the xanthophyll cycle resulting in seed yield increases of up to 33%.

Ermakova M, Danila FR, Furbank RT, Von Caemmerer S (2019) On the road to C_4 rice: advances and perspectives, *Plant J* 19, 940–950. DOI:10.1111/tpj.14562

Ermakova M et al (2021) Installation of C_4 photosynthetic pathway enzymes in rice using a single construct, *Plant Biotechnol J* 19, 575–588. DOI: https://doi.org/10.1111/pbi.13487

Two linked papers describing how C_4 photosynthesis might be transferred to rice.

Gastaldo RA (2019) Plants escaped an ancient mass extinction, *Nature* 567, 38–39. DOI: https://doi.org/10.1038/d41586-019-00744-3

Evidence that plants might be less susceptible to mass extinction events than animals.

Hibberd JM, Furbank RT (2016) Fifty years of C4 photosynthesis, *Nature* 538, 177–179. DOI: https://doi.org/10.1038/538177b

Useful review of research into C4 photosynthesis.

Huisman R, Geurts R (2020) A roadmap toward engineered nitrogen-fixing nodule symbiosis, *Plant Commun* 1, 100019. DOI: https://doi.org/10.1016/j.xplc.2019.100019

A possible route to engineering nitrogen fixation into new plant species.

Turvey ST, Crees JJ (2019) Extinction in the anthropocene, *Curr Biol* 7, R982–R986. DOI: https://doi.org/10.1016/j.cub.2019.07.040

Useful primer on our current mass extinction event, albeit mostly animal focused.

Wang Q et al (2020) Molecularly engineered photocatalyst sheet for scalable solar formate production from carbon dioxide and water, *Nature Energy* 5, 703–710. DOI: https://doi.org/10.1038/s41560-020-0678-6

Using biomimetics to create artificial photosynthesis.

Zhu XG, Ort DR, Parry MA, von Caemmerer S (2019) A wish list for synthetic biology in photosynthesis research, *J Exp Bot* 71, 2219–2225. DOI: https://doi.org/10.1093/jxb/eraa075
Some of the latest options for using synthetic biology to improve photosynthesis.

Discussion questions

7.1 Describe the major changes in the contemporary terrestrial flora that have been caused by human activities over the past 11,500 years.

7.2 Outline two ways in which biotechnology could be used to improve photosynthesis and CO_2 fixation.

7.3 Discuss the major challenges facing plant life and how these might be tackled.

GLOSSARY

Abiotic stresses non-living, environmental factors potentially harmful to growth or development of an organism, such as drought, salinity, and mineral deficiency. See *Biotic stresses.*

Algae highly diverse group of mostly aquatic photosynthetic eukaryotes, nearly all with plastids originally derived from an endosymbiotic cyanobacterium; see *Primary algae* and *Secondary algae.*

Aerobic organism or environment containing free oxygen.

Anthropocene see *Holocene.*

Anthropogenic events or processes caused by humans.

Amyloplast type of colourless plastid responsible for starch storage, found in starchy tissues such as cereal grains and tubers.

Anaerobic organism or environment lacking free oxygen.

Angiosperms seed-bearing, flowering plants making up the largest and most diverse extant group of higher plants. Divided into the monocots, basal dicots, and eudicots, and containing many species of major economic importance.

Anoxygenic photosynthesis use of light energy to oxidize various electron donating species, such as hydrogen gas (H_2), hydrogen sulphide (H_2S) ferrous iron (Fe^{2+}), thiosulphate ($S_2O_3^{2}$) or sulphur (S_8), in order to generate protons and electrons without the release of oxygen.

Apicoplast relict non-photosynthetic plastid found in most Apicomplexans.

Archaea one of the two fundamental domains of life. The other domain is the Bacteria. Eukaryotes originated as a branch of the Asgard group of archaea.

Archean the second eon in geological time, running from 3.9 to 2.5 Ga, during which several groups of photosynthetic bacteria evolved, including cyanobacteria.

Archaeplastida monophyletic eukaryotic group resulting from the primary cyanobacterial endosymbiosis. It includes all of the primary algae (red, green, and glaucophytes) and the land plants.

Assimilates see *Assimilation.*

Assimilation conversion via photosynthesis of simple molecules such as minerals and CO_2 into more complex molecules and structures such as carbohydrates, proteins, cells, and tissues. Assimilates such as sucrose and amino acids are transported via the *phloem.*

Autotroph organism capable of synthesizing complex biomolecules from simple inorganic precursors using either light (photoautotroph) or chemical (chemoautotroph) energy.

Bacteria one of the two fundamental Domains of life; the other Domain is the Archaea.

Biofuels renewable fuels derived from recently living organisms as alternatives to non-renewable fossil fuels. Examples include bioethanol from cereals and biodiesel from oil crops.

Biomass mass, often expressed as tonnes of carbon, of a biological species or group of species in a given area.

Biomimetic using non-living materials to mimic biological processes. Examples include aircraft wings based on bird wings, and artificial photosynthetic systems.

Biota the totality of living organisms.

Biotic stresses negative effects caused by other living organisms, such as parasites, pathogens, and herbivores. See *Abiotic stresses.*

Boreal subarctic ecosystem between latitude 50° and 70° N, often characterized by coniferous forests.

Bryophytes earliest diverging lineages of extant land plants consisting of small, non-vascular species with a haploid adult generation; includes mosses, liverworts, and hornworts.

CAM (Crassulacean Acid Metabolism) pathway found in some arid zone plants involving CO_2 fixation during the night, hence minimizing water evaporation during the day.

C3 Photosynthesis the most common mechanism for CO_2 fixation in plants where the initial carboxylation product is the C3 compound, 3-phosphoglycerate.

C4 Photosynthesis alternative mechanism for CO_2 fixation in plants where the initial carboxylation products are the C4 compounds, oxaloacetate, and malate.

CASH (Cryptophyte, Alveolate, Stramenopile, and Haptophyte) supergroup of eukaryotic protists

Cereal cultivated species of the grass family, Poaceae, with relatively large starch-rich grains or caryopses. Important crops include maize, rice, wheat, and barley.

Charge separation process in photosynthetic reaction centres whereby light energy leads to the production of positively and negatively charged groups which must then be rapidly separated in order to avoid their recombination and the loss of potentially useful energy.

Charophytes group of green algae whose ancestral lineage gave rise to land plants or Embryophytes.

Chemoautotroph organism able to use chemical energy to synthesize complex organic molecules.

Chlorophyll class pigments found in photosynthetic organisms. The primary pigment molecule used in oxygenic photosynthesis by cyanobacteria, algae and plants, but can also be found in many anoxygenic phototrophs.

Chlorophytes one of the two groups that constitute the Viridiplantae (*Streptophytes* are the other group), Chlorophytes include the major groups of green algae.

Chloroplast most common type of plastid. Site of photosynthetic light and dark reactions which occur respectively on thylakoid membranes and stroma.

Chromoplasts type of pigmented plastid found in coloured plant tissues such as flower petals and fruits; pigments are stored in lipid droplets called elaioplasts.

Clade group of organisms sharing a common ancestor. See *monophyletic*.

CO_2 fixation process whereby CO_2 from the atmosphere or dissolved in water is converted into organic carbon compounds.

Cyanobacteria class of photosynthetic prokaryotes that use two chlorophyll-containing reaction centres to carry out oxygenic photosynthesis using water as a reductant. All plastids in algae and plants are descended from cyanobacteria.

Cytochrome b_6f complex protein complex linking photosystems I and II to enable electron transport from water to NADPH.

Domain the most fundamental group for classification of biological organisms. The two Domains are Bacteria and Archaea, with eukaryotes arising from deep within the archaea.

Elaioplast non-photosynthetic lipid-enriched plastid involved in synthesis and storage of several types of lipidic metabolites.

Electron transport series of redox reactions in which electrons are passed rapidly along a chain of acceptors in processes such as respiration and photosynthesis.

Embryophytes a monophyletic group including all of the non-algal land plants.

Endosymbiosis form of symbiosis where one partner, the endosymbiont is fully engulfed by and resides within another partner, the host. In some cases the endosymbiont maintains its autonomy and can survive outside the host cell. More commonly, the endosymbiont undergoes selective gene loss and eventually loses autonomy. Two major eukaryotic organelles, mitochondria and plastids, resulted from ancient endosymbiotic events with bacterial cells at least 2.4 Ga. See *Primary, Secondary,* and *Tertiary Endosymbiosis*.

Eon largest division of geological time. Four eons are recognized: Hadean (4.55 to 3.9 Ga), Archean (3.9 to 2.5 Ga), Proterozoic (2.5 Ga to 540 Ma) and Phanerozoic (540 Ma to present).

Etioplast colourless chloroplasts found in dark-grown plant tissues that have not produced pigments or thylakoid membranes but which contain the precursors in the form of semi-crystalline membrane lattices.

Eudicot one of the three main divisions making up 75% of angiosperm species, the other two being the basal angiosperms and monocots.

Eukaryotes secondary but highly important branch of the primary Domain, Archea.

Extant a species that is still present and alive in contrast to one that is extinct.

Facultative process of which an organism is capable but is not essential for its existence. See *Obligate*.

Ga (giga annum or billion years) measure of time used for the earlier periods of earth history, typically used for events older than 1 Ga. See *Ma*.

Genome the genetic complement of an organism, including functional genes and non-coding DNA. The principal genome of eukaryotes resides in the nucleus but much smaller genomes are present in mitochondria and plastids.

Genomics description and analysis of genomes.

Glaucophytes small group of primary algae. Red and green algae are the other primary algae. See *Archaeplastida*.

GOE1 (1st Great Oxygenation Event) increase in global oxygen from near zero to about 1% of present levels during the early Proterozoic, ca 2.4 Ga.

GOE2 (2nd Great Oxygenation Event) increase in global oxygen levels from about 1% to close to present levels during the late Proterozoic, ca 0.8 Ga.

Grana stacked thylakoid membranes inside chloroplasts that house most of the photosynthetic apparatus.

Green algae largest and most diverse group in the *Archaeplastida*. Together with land plants they make up the *Viridiplantae*.

Gymnosperms major group of seed bearing but non-flowering higher plants most of which are coniferous trees, including pines, firs, spruces, and cycads.

Heterotroph organism that can only obtain complex organic molecules from other living or dead organisms.

HGT (Horizontal Gene Transfer) naturally occurring transfer of DNA segments, including entire genes or groups of genes, from one species to another unrelated species. Many genes have been transferred between plants, animals, viruses, and bacteria, and this is an evolutionarily important mechanism for generating new genetic diversity.

Holocene relatively warm and stable period from about 11,500 years ago until now, when humans invented agriculture and had an increasingly profound effect on global flora and fauna. 'Anthropocene' is a more recent term used to cover the same period, although sometimes it only describes the industrial era of the last two centuries.

Kleptoplasty temporary acquisition of plastids by a heterotrophic cell following the ingestion of an algal cell.

Light harvesting photosynthetic process for the collection of a relatively wide range of light wavelengths and the channelling of the photons into the reaction centre complexes.

Lignin a complex heterogeneous polymer based on phenylpropane units that is the major structural component of wood.

LECA (Last Eukaryotic Common Ancestor) earliest eukaryotic organism that is the ancestor to all extant eukaryotes. LECA was unlikely to have been the first eukaryotic cell but it is the only one to have descendants today.

Lipid substance soluble in organic solvents, eg phospho- and glycolipids, sterols, waxes, and triacylglycerols. In plants, lipids function as the matrix of biological membranes, as storage reserves, and as hormone-like mediators.

LUCA (Last Universal Common Ancestor) earliest form of life that is the ancestor to all extant organisms. LUCA was unlikely to have been the first life form but it is the only one to have descendants today.

Lumen aqueous phase inside the thylakoid membrane sacs of chloroplasts.

Ma (mega annum or million years) measure of time before present, typically used for events from 1 to 1,000 Ma. See *Ga*.

Mixotroph organism that use both *autotrophic* and *heterotrophic* modes of nutrition.

Molecular clock method of genetic analysis used to estimate evolutionary rates and timescales by comparing DNA or protein sequence data.

Monocots major group of flowering plants or angiosperms that includes grasses and palms plus many important food crops such as wheat, rice, and maize.

Monophyletic a group of species or *clade*, that shares a single common ancestor and also includes all of the descendants of that common ancestor.

Montane higher altitude ecosystem, often characterized by subarctic vegetation.

Motile capable of autonomous movement using metabolic energy. See *sessile*.

Multicellular organism consisting of numerous cells, normally containing several distinct tissue types.

Mycorrhizae fungal species associated with the roots of most higher plants that receive assimilates from plants in return for soil-borne nutrients and water.

Nitrogenase large prokaryotic enzymatic complex able to fix atmospheric N_2 into ammonia, which is then converted into nitrates. Both ammonia and nitrates can be taken by plants for biosynthesis of organic nitrogenous compounds such as amino and nucleic acids.

NADPH nicotinamide adenine dinucleotide phosphate (reduced), biological carrier of reducing equivalents.

Nitroplast see *Spheroid bodies*.

Obligate process or lifestyle that is essential for growth of an organism. See *Facultative*.

Oomycete superficially fungus-like organisms that include many virulent plant pathogens. Unlike true fungi, oomycetes have cellulose cell walls and may be derived from algae that lost their plastids after assuming a heterotrophic lifestyle.

Oxygenic photosynthesis use of light energy to split water molecules (H_2O) into protons (H^+) and electrons (e^-) with the release of oxygen gas (O_2) as a side product.

Ozone layer stratospheric ozone layer (15–30 kilometres high) that absorbs high-energy solar UV radiation that would otherwise be harmful to most terrestrial organisms. In contrast, ground level (tropospheric) ozone can be a dangerous pollutant damaging plants and reducing crop yields.

Phanerozoic the fourth eon in geological time, running from 540 Ma to the present, during which land plants evolved to dominate terrestrial ecosystems.

Phloem active transport system in vascular plants for movement of *assimilates*.

Photoautotroph organism able to use light energy to drive ATP synthesis, either via electron transport chains or direct proton pumping.

Photosynthetically active radiation (PAR) light energy between 400 and 700 nm that is the optimal range for plant photosynthesis. Many algae and bacteria have wider PAR ranges.

Photosystem large thylakoid membrane-bound pigment protein complex, also known as *photochemical reaction centre*. Plants, algae, and cyanobacteria contain two photosystems, I and II, that act in series in oxygenic photosynthesis.

Phototroph organism able to use light energy to drive ATP synthesis, either via electron transport chains or direct proton pumping. Some phototrophs, such as archaea are unable to fix CO_2 and are therefore incapable of photosynthesis.

Phycobilisomes large pigment-protein arrays used for light harvesting. Found on the thylakoids of cyanobacteria, red algae, and glaucophytes but absent from green algae and land plants.

Plants informal synonym for *Embryophytes* or land plants.

Plasmodesmata membrane-lined channels linking adjacent cells via their cell walls enabling the regulated movement of molecules throughout much of the plant body.

Plastids class of DNA-containing, semi-autonomous organelles ultimately descended from cyanobacterial endosymbionts. Examples include proplastids, chloroplasts, chromoplasts, and amyloplasts.

Primary algae major and possibly monophyletic group in the Archaeplastida comprising the red, green, and glaucophyte algae, all of which contain plastids originating from a single endosymbiotic event between a heterotrophic eukaryote and a cyanobacterium. Land plants are derived from the Streptophyte group of green algae.

Primary Endosymbiosis uptake of a cyanobacterial cell by a phagotrophic eukaryotic cell and conversion of the endosymbiont into a primary plastid organelle.

Primary production biological process for the synthesis of organic compounds from CO_2, principally via photoautotrophy, but can also occur via chemoautotrophy. See *Autotrophy*.

Prokaryotes unicellular or colonial organisms that lack a nucleus and most organelles. The two prokaryotic groups, bacteria and archaea, constitute the primary domains of life.

Proplastid precursor stage in plastid life cycle, proplastids can develop into several alternative types of plastid as specified by their host cell or tissue.

Proterozoic the third Eon in geological time, running from 2.5 Ga to 540 Ma, during which oxygenic cyanobacteria and algae created an oxygen-rich atmosphere that enabled aerobic organisms to become the dominant life forms on Earth.

Quinones class of aromatic compounds with numerous biological roles, such as the photosynthetic electron carrier, plastoquinone.

Reaction Centres protein complexes that are the sites of the light reactions of photosynthesis, also known as photosystems.

Reactive oxygen species (ROS) oxygen radicals and non-radicals formed during normal metabolic processes such as respiration and photosynthesis, or as a consequence of many forms of stress. ROS can damage biomolecules leading to injury, premature senescence, or death. Examples include: hydroxyl (OH), peroxyl (ROO), superoxide (O_2), and alkoxyl (RO) radicals; and the non-radical intermediates singlet oxygen ($^1O_2$), hydrogen peroxide (H_2O_2), and ozone (O_3).

Red algae (Rhodophytes) one of three major groups in the *Archaeplastida*.

Reductive pentose phosphate (RPP) pathway also known as the Calvin or Calvin-Benson pathway or cycle, this is the most important pathway for the fixation of CO_2 via *rubisco* and the generation of simple carbohydrates using NADPH and ATP generated during the light reactions of photosynthesis.

Rhizosphere soil region in the vicinity of plant roots and its associated flora, such as plant-associated bacteria and *mycorhizae*.

ROS see *Reactive oxygen species*.

Rubisco RibUlose BISphosphate Carboxylase/Oxygenase, the principal enzyme responsible for fixation of CO_2 into complex organic molecules during photosynthesis, see also *RPP* pathway.

Secondary algae highly diverse paraphyletic groups of algae that contain plastids derived from a primary algal endosymbiont.

Secondary Endosymbiosis uptake of a primary algal cell by a phagotrophic eukaryotic cell and it's conversion into a secondary plastid organelle.

Sessile organism that remains fixed in one place. See *motile*.

SNARE (Soluble *N*-ethylmaleimide-sensitive factor Attachment protein REceptor): class of protein involved in endomembrane fusion in eukaryotes and some archaea.

Spheroid bodies permanent intracellular endosymbionts of some diatoms, that are evolving towards becoming a nitrogen-fixing organelle.

Stomata pores in aerial tissues of most land plants that mediate gas and water exchange.

Stomatophyte hypothetical ancestor of the *Embryophytes* characterized by presence of stomata.

Stramenopiles also called heterokonts, a major eukaryotic group most of which are secondary algae, such as diatoms and kelp, or their non-photosynthetic descendants, such as the parasitic oomycetes.

Streptophytes one of the two groups that constitute the *Viridiplantae* (*Chlorophytes* are the other group), Streptophytes include the green algal *Charophytes* and the *Embryophytes* or land plants.

Stroma main aqueous phase in plastids that houses the CO_2 fixation and starch synthesizing enzymes.

Stromatolites macroscopic layered formations created by consortia of prokaryotes, including photosynthetic bacteria, and often found in shallow water with fossil forms dating from 3.7 Ga or earlier.

Superoxide dismutase (SOD) enzyme able to convert the highly reactive superoxide radical into molecular oxygen.

Symbiosis persistent association of two unrelated organisms termed symbionts. Although mostly used for mutually beneficial relationships, symbiosis can also include more one-sided interactions such as parasitism.

Syntrophy form of symbiosis frequently found in microbial ecosystems involving shared metabolism between two or more species.

Terpenoid class of compound made up of one or more isoprene units that includes carotenoids, many plant hormones and important herbicides such as pyrethrins

Tertiary Endosymbiosis uptake of a secondary algal cell by a phagotrophic eukaryotic cell and its conversion into a tertiary plastid organelle.

Thermophilic organism that grows optimally at high temperatures, typically above 50 °C and in some cases as high as 122 °C.

Thylakoids system of stacked and unstacked flattened membrane sacs in chloroplasts housing pigment-protein and ATP-forming complexes involved in photosynthetic light reactions.

Tracheophytes see *vascular plants*.

Tree of Life term first used in a biological context by Charles Darwin. Examples are normally depicted as phylogenetic trees showing the interrelatedness of extant and extinct organisms based on shared genetic and morphological features.

Troposphere lowest level of the atmosphere extending to a height of between 6 and 18 km above the surface. Tropospheric ozone can be toxic to life, whereas stratospheric ozone at 7 to 18 km above the surface provides a vital shield against potentially damaging high-energy solar radiation.

Unicellular organism consisting of a single cell.

Vascular plants land plants belonging to the *pteridophytes* and *spermatophytes* having a transport system made up of *xylem* and *phloem* tissues.

Viridiplantae (aka Chloroplastida) group within the Archaeplastida that comprises the green algae and land plants (Embryophytes).

Xylem passive transport system in vascular plants for movement of water and minerals.

Z-scheme shorthand name for linear electron transport pathway of oxygenic photosynthesis.

SUBJECT INDEX